# 60 DAYS *to a* GREENER LIFE

## Ease Eco-anxiety Through Joyful Daily Action

## HEATHER WHITE

HARPER
HORIZON

# INTRODUCTION

S cientists warn that we have less than six years before the climate crisis reaches a point of no return.[1] That's the inconvenient and unsettling truth. Coming to terms with the enormity of the climate emergency can be overwhelming. We can lean into this vulnerability, experience the full range of emotions that come with it, and also celebrate the joy of climate action.

Given the recent extreme weather around the globe, it's little wonder people are experiencing "eco-anxiety," also known as "climate anxiety," intense emotional stress about climate change. This is especially true for Generation Z (those born after 1997), who acutely worry about the crisis. A recent global survey of ten thousand students found that nearly half of young people say that eco-anxiety is interfering with their daily life. One out of four don't want to have children of their own because they are so worried about the future.[2] Gen Z also understands that climate policy must center racial, economic, and social justice, known as "climate justice," as we build a regenerative, hopeful future.[3]

We all know that one person can't solve the climate crisis and that we need comprehensive policy and market action. My teens have made fun of me for skipping the straw. "Like that's really going to help, Mom," they've said with an eye roll. But the fact is that millions of people around the world skipping the straw, saying no to single-use plastic, and shining a bright light on plastic pollution made a difference. These actions spurred the United Nations and 175 countries to agree to legally binding treaty negotiations to end global plastic pollution.[4]

You matter.

Being intentional with how we live each day—from what we eat to how we connect to our spirit, our community, and the earth—will give us a shot at a greener, healthier future by *changing the culture*. Most of all, celebrating a daily practice of sustainability is uplifting and fun. That's where this book comes in.

Each day you'll read about a topic that impacts our planet. I'll provide you with information so you can understand how you can make a difference. Sometimes there will be a call to action and other times I'll ask you to read more information on a topic. Every day will be different. After reading, you'll have a better understanding on how you can impact the world and the steps you can take to live a greener, more eco-friendly life.

Be kind to yourself and remember that you're a product of a system that shifted disposal and energy emissions to you, the consumer, instead of the manufacturers who made all this stuff. The daily ritual of living a greener life is about intention and cultural change, not blame or shame. Climate activism isn't about perfection—it's about purpose, personal growth, and daily action.

Yours in partnership for a healthier, greener, more equitable world,

*Heather White*

## DAY 1
# Go Outside & Find the Awe

Deciding to live a greener life can be as easy as writer Anne Lamott's famous advice: "Go Outside. Look Up. The Secret of Life."[5] Of course, going outside to look up for five minutes isn't going to solve the climate emergency. However, that five minutes can restore you. It can inspire you. It can change your perspective. It can bring you hope and awe. Even better, the action of going outside and observing can make you look forward to the next green step you're ready to take. Who knows? You might even discover your own version of Lamott's famous advice with a few of these small but compoundable actions.

The challenge for most of us is the basic but tough question: *How do I start? It's important to recognize that the small steps you take matter. Going outside is a great way to reduce feelings of stress and anger. Spending time in nature is linked to learning benefits and improvements in mental health and emotional well-being.* Key upsides of spending time outside include improved physical fitness, more vitamin D, enhanced sleep, decreased anxiety, lowered stress hormone levels, reduced ADHD symptoms, and increased creativity and self-esteem.[6] *The feeling of awe can calm negative self-talk and increase your sense of purpose.[7] Today, go outside and breathe in the fresh air. Commit to sitting outside on your porch or deck, taking a walk, or riding a bike.*

*Appreciate the planet and the feeling of awe that nature provides.*

Did you have a favorite outdoor spot to visit when you were younger? Describe what it was like. What made it special? How did you feel in that space? What did you see, hear, smell, touch, or taste?

## DAY 2

# Understand Eco-anxiety & Feel the Feels

The term *eco-anxiety*, also called "climate anxiety," is a relatively new trend that many doctors and psychologists are witnessing. In 2017, the American Psychological Association recognized eco-anxiety as a "chronic fear of environmental doom."[8] The Climate Psychology Alliance formed to train mental health professionals to identify and treat eco-anxiety. A recent survey of child psychiatrists in the United Kingdom discovered that 50 percent had clients who suffered from it.

For Gen Z, "eco-anxiety" has three aspects. First, *children are suffering from generalized anxiety in greater numbers*. Each child experiences anxiety differently, but the statistics are alarming. The National Institute of Health indicates that 30 percent of American teens suffer from anxiety.[9] In 2021, the United States surgeon general issued an official public health advisory on the issue of teen depression and anxiety.[10]

Second, *Gen Z is the loneliest generation*. More screen time and less in-person interaction mean a sense of isolation for them, even before the pandemic. In the 2022 Cigna Loneliness Index, eight out of ten Gen Zers experienced chronic loneliness compared to four out of ten Baby Boomers.[11] Today's young people are lonelier than the elderly.

Third is the *hyperawareness of the climate crisis*. Gen Z is asking what the future will look like, where they will live, and what their experience will be like on this planet. In a 2020 survey by the US Conference on Mayors, 80 percent of Gen Z agrees that "climate change is a major threat to life on earth"; one in four have taken direct action on climate change, and by three to one, Gen Z believes "the climate crisis warrants bold action."[12] The extreme weather events we've experienced and the inequities around the climate crisis are lot to bear.

In today's culture we go to extraordinary lengths to help children avoid discomfort in order to ease our own anxiety about our children's pain. Yet recent research shows that kids with clinical anxiety have to be part of the solution in dealing with their stress. Fixing it for them doesn't help.[13]

 Do you know a Gen Zer with anxiety? If yes, have you talked with and listened to the young people in your life about their anxiety and their thoughts about the future they're inheriting from us?

# Learn about the Law of Simplicity & Consistency

Small actions alone will not solve the climate crisis and neither will the actions of one person. Only one hundred companies are responsible for 71 percent of greenhouse gas emissions since 1988. And only 8 percent of plastics are recycled.[14] We need substantial policy and market solutions, but sometimes we overlook an important point. Individual action can create ***culture change*** so these comprehensive strategies work.

Tiny changes create momentum. You can find a ton of research on this phenomenon in leadership books, but my favorite resource is James Clear's *Atomic Habits*. He reminds us that "we often dismiss small changes because they don't seem to matter very much in the moment."[15] But if we create systems that allow for small changes, their impact compounds over time. Our daily actions make up our lives. If we focus on tiny, consistent actions, we can make time itself a powerful agent for transformation of our individual and collective experiences.

To make a habit repeatable, you need it to be obvious, attractive, easy, and fun.[16] Another way to think about creating a new habit is focusing on cue, action, and reward, as Charles Duhigg writes about in *The Power of Habit*. This means that something reminds you to take action (the "cue"), you take the action, and then your brain is rewarded for that action. Naming the reward is powerful enough to create a new habit.[17] In the context of living a greener life, the feeling of positive contribution to the planet and your community can be the reward.

Think about your daily routine. What are you already doing that might be considered eco-friendly? What other simple and consistent steps can you take? If you're stuck, here are a few suggestions:

Reduce your food waste by planning your meals and eating your leftovers.

Unplug your appliances and your cell phone charger when you aren't using them to prevent them from draining energy, and get a programmable thermostat.

Rid yourself of single-use plastic by bringing a reusable mug to the coffee shop, skipping the bottled water, bringing your own bags to the store, packing your lunch in reusable containers, or asking your favorite takeout place to move to compostable packaging.

# Find Your Climate Why
# & Show Gratitude

**W**hen I was growing up and my grandma would talk about my dad, she'd say, "That's when you were just a twinkle in his eye." She meant that she was talking about a long time ago, way before I was born. When I discuss the mindset of compassion and resilience in this book or with others, I encourage them to create their own "why" for climate action. My why is all about the twinkles in my eye and my kids' eyes. I want to be an *awesome ancestor*.

My paternal grandma, Wanda, left high school at seventeen to get married and have kids. When she was in her mid-forties, she returned to school to get her high school diploma. No one encouraged her, but she did it because she believed in the power of education. My great-grandma on my mother's side, Lovie Jane, didn't know how to read or write. Granted, she was born in the 1880s in Appalachia. Lovie Jane signed her name with an *X*, as my mom recalls. And here I am, writing a book. Let that sink in.

At random moments, I think about my grandparents and the generations of change agents I never knew or met who have helped me. They made advances in science, secured our right to vote, and created educational opportunities. Then there were those whose survival itself was heroic, those who kept going when it seemed impossible. That desire to honor my past and pay it forward to future generations is a strong motivator for me. This intergenerational duty moves across cultures, too, whether it be the concept of the Seven Generations in Native American culture or the Ghanaian concept of "Sankofa," a turning back and reaching for knowledge in the past.

We need to act fast. The International Panel on Climate Change says that we have until 2030 to make enormous progress on climate change, or we may reach a point of no return. Even in the polarized political world we currently live in, we must change the culture. Part of that change is individual action, but it's also believing that change is possible.

Today, I want you to think about your climate why. Why did you pick up this book?

Next, think about your ancestors and the advances they made possible. Is there someone in your family or friend circle that inspires you to make a difference for the future? Who is this person? What did they accomplish?

Think of someone you can thank for their contributions. Is there a teacher who believed in change? Is there a coach that reinforced individual action? Pick up the phone or write a letter (Yes, seriously! Use a stamp.) to share your gratitude.

# Visualize 2030

What does a world look like in which we value sustainability, have reduced global carbon pollution, and flourish in a regenerative economy? What does it feel like? How does personally valuing climate action and service work? How do equity and justice play a role in the future vision? Envision what a regenerative, positive 2030 would look like: Rooftop gardens. Solar wind farms. Clean water and clean air for all. Green buildings. Mass transit. Urban parks. Expansive nature preserves.

Be creative. Take out a piece of paper or make space in a journal. Draw it. Write it. Use word imagery. Make it as specific and as hopeful as you can.

What will your 2030 life be like? Think about your family and friends in 2030. Where will you live? What will you do for a living? How do you spend your downtime? What does a thriving community look like to you? What do you eat? Where does your food come from? How do you interact with others? What role does technology play? What role do you want it to play? What role does nature play? What role should it play?

Today is all about visualizing 2030 so you can see it, believe, and plan. Write down the steps you need to take so you can achieve a more joyful, eco-friendly life. Seeing your action plan will make things feel more tangible and doable.

Visit www.onegreenthing.org/2030visualization to listen or download the visualization exercise.

# DAY 6
# Start with Connection

**F**rom Times Square to the TED stage, tech entrepreneur Asher Jay's artwork inspires people across the planet to respect and protect wildlife. A world traveler, photographer, and artist, Asher was named a NatGeo Explorer by National Geographic.

Asher calls herself a "creative conservationist" who loved environmental science as a kid. "I read David Attenborough as a first or second grader, with a dictionary by my side. I was obsessed with the natural world" and "highly sensitive to global things, particularly the plight of the earth." One day she was watching a program about chimpanzees being experimented on and cried so hard that her parents intervened. "No more watching TV, no more books about wildlife," declared her father.

"I was way too emotional and consumed by the harm to the

earth, from a place of victimhood and sorrow," she says. Asher had to find a different way. She was modeling at the time, and decided to go into fashion and studied at Parsons.

Then the 2010 BP oil spill happened. When she saw dolphins dying and other wildlife covered in thick oil, environmental activism pulled her back. She recalled that the cause of the spill was an improperly installed drilling rig. "To think that a few bolts could result in the destruction of so much life broke my whole reality apart." When BP started spraying dispersants to break up the oil, the efforts to control the damage made the impact worse. "I knew I had to be the change. In 2009, I lost my dad. He had agency and chose his course of treatment. These dolphins, though, did not. Twenty thousand marine mammals dying was unacceptable. Every cell in my body was telling me I had to do something."

Asher channeled her pain into action. "I chased environmental thought leaders I wanted to meet and hopped in cabs to the airport with them to learn more. I wanted answers." Soon she was on TV talking about the BP spill and plunged headfirst into conservation, art, and the climate movement.

What gives Asher hope is her transformation from anger to acceptance and a firm belief that it'll be okay. Her advice is to start with connection. "Find a moment to consider a living thing outside yourself. Spend time with it. It could be a grasshopper that you notice or an inside house plant," stated Asher. "It's not a commodity. If you make that genuine connection outside of yourself, you can create space to show up more."

Answer Asher's question: what is one thing outside of yourself that you can make a genuine connection with? How can you show up more for that thing? If you're stuck, here are a few examples: a pet, a favorite tree, or a local stream.

How can you share this thing with others so together more good can be created? If you're stuck, using the examples above, here's how you can invite a friend:

Pet: Go walk your pet with a friend or make a plan to volunteer at an animal shelter.

Tree: Grab a guide and ask your friend to identify some local trees with you.

Local stream or park: Invite friends from different generations for a hike or walk to your local stream or park.

## DAY 7
# Define Climate Change & Global Warming

*Climate change* technically refers to changes in temperature and weather patterns over a long-term period. *Global warming* is the increase in temperature of the earth's atmosphere from the greenhouse effect, where the sun's rays continue to heat the earth as heavy greenhouse gases act as a blanket and trap other lighter gases from escaping. This leads to a warming of the earth, like what happens in a typical greenhouse. The major greenhouse gases are carbon dioxide, methane, nitrous oxide, water vapor, and chlorofluorocarbons. The terms *global warming* and *climate change* are now often used synonymously.[18] Here are three key facts regarding climate change:

- According to NASA, humans have increased carbon dioxide in the atmosphere by 47 percent since the Industrial Revolution.[19]
- There is greater than 99 percent scientific consensus that human activity—mostly from energy choices that burn oil, coal, and gas and emit carbon dioxide—is the cause of the climate emergency.[20]
- On August 9, 2021, the Intergovernmental Panel on Climate Change's report, which was approved by 195 countries and based on 14,000 studies, concluded that the evidence that human activity causes global warming is "unequivocal."[21]

Even though a solid majority of 72 percent of adult Americans have accepted the science that climate change is real and a global threat, that number falls short of the 99 percent of scientists who accept this reality.[22] Compare this to 80 percent of Gen Zers, who believe that climate change is "a major threat to human life as we know it."[23] Yet people who support climate action tend to **underestimate** how many people agree with them by two to one.[24]

Do you feel like you should be doing more? Or that you're using too much plastic? Not recycling the right stuff? Don't have a hybrid car? Be compassionate to yourself. Understand that the whole concept of the individual carbon footprint was a public relations gimmick by the oil and gas industry. The goal was to shift the burden to consumers, instead of the companies who are responsible for this crisis. Don't fall for this spin.[25] You didn't cause this mess, but you are part of the solution. The daily ritual of greener living drives progress.

Do you have family members or friends who don't think the climate crisis is a problem? If yes, write down three things you'd like to discuss—compassionately and respectfully—with friends who aren't sure. If no, write down three ways you can share more about your worries about and hope for climate solutions.

# DAY 8
# Know the Facts

Here are a few facts to share with any family or friends who may remain skeptical about what we're experiencing. Even though there have been climate cooling and warming periods throughout the earth's history, climate change is "happening 20 to 100 times faster than the most rapid changes in climate history."[26] And while weather may fluctuate from day to day or area to area, climate refers to a longer time period. On average, our climate is on track to be three degrees Celsius hotter by 2100, and over the past century, it's already warmed by one degree Celsius. The largest amount of carbon pollution comes from the energy and transportation sectors. We know fossil fuels are the major contributor because their carbon pollution has a unique carbon dioxide fingerprint, called delta C thirteen, which has dramatically increased since the 1880s.[27]

If some of your family members or climate-curious friends don't believe the international climate scientists, then they might believe the oil and gas industry's own experts. As early as 1959, internal documents show that the American Petroleum Institute (API) stated that oil and gas drilling contributed to global warming. Another example is a 1980 internal document, which details a meeting with oil company members of API and John Laurmann of Stanford University, whom API hired to advise them on the latest in atmospheric research. Laurmann warned that if the world didn't shift to clean energy, then the following would happen:[28]

- 1°C rise (2005): Barely noticeable
- 2.5°C rise (2038): Major economic consequences, strong regional dependence
- 5°C rise (2067): Globally catastrophic effects

Check out Ben Franta's TED Talk from July 2021, where he explains that this scientist told API in 1980 that a warming of

2.5 degrees Celsius could "bring world economic growth to a halt" and suggested that "avoiding the predicted outcomes would require prompt action, since the adoption of non-fossil-fuel energy sources would likely require decades to accomplish."[29] The award-winning exposé from *Inside Climate News* also leaves no doubt that, similar to the tobacco companies and asbestos manufacturers, Big Oil knew its products were the main contributors to global warming and that the impacts would be potentially catastrophic.[30]

 Watch Ben Franta's TED Talk.

# DAY 9
# Stop Buying So Much Stuff

**M**ost of us are drowning in stuff—clothes, plastic toys, old or broken appliances, used electronics, papers, memorabilia, and general clutter. This chapter isn't about shame. Overbuying has become part of modern American culture, but we can change what we buy and why by being intentional. Groups like the Buy Nothing

Project encourage us to share, reuse, or recycle items we no longer need but may be useful to others. Consider these eye-opening stats:[31]

- The average U.S. household contains 300,000 things.
- According to the National Association of Professional Organizers, Americans don't use 80 percent of the items they own.
- One in ten Americans rent storage units to store stuff they don't use.
- If you have a room in your house you can't use because it's filled with clutter, you are not alone. One in seven Americans experience this problem.
- A recent survey showed that one out of three Americans has experienced "extreme anxiety" over clutter, and two-thirds of us know we own too much stuff.
- The Story of Stuff project found that the average American's waste has doubled over the past thirty years to four and a half pounds of garbage a day.
- When we take out the trash, every garbage can we put out equals seventy garbage cans of upstream waste to make the stuff we threw away.

Here are some strategies to manage what you buy and the stuff you own:

- Celebrate "Buy Nothing Day." Try to make it through a day without actively buying anything. See if you can make this a monthly or weekly practice.

- Declutter and resell or donate what you can't use.
- Make a practice of embracing the concept that less is more.
- Host a community garage sale.
- Set aside a weekend day and digitize papers, old photos, and other sentimental items.

 Track what you buy for a week. How can you streamline to save money and reduce your impact on the environment?

Join the Buy Nothing community in your area to exchange items you no longer want or to find free items you may need.

## DAY 10
# Choose Sustainable Fashion & Thrift

S ustainable fashion is clothing that is designed, manufactured, distributed, and used in ways that are environmentally friendly. Sometimes these products can be more expensive. However, the ultimate price tag is the damage fast-fashion brands are placing on the planet. Fast fashion is intentionally designed for a cheap price and viewed by customers as disposable. Over the past twenty years, people buy on average 60 percent more clothing.[32] Yet, according to the World Bank, 10 percent of global carbon emissions come from the fashion industry, "more than all international flights and maritime shipping combined."[33] Nearly 90 percent of fiber used in fashion ends up in a landfill or is incinerated. The industry contributes one-fifth of the world's water pollution from fabric treatment and the dyeing process.[34]

Most clothes are made out of plastic fibers, creating a microplastic disaster. There are also thousands of harmful chemicals used in the textile mills around the world—which are dangerous to the environment and the workers. Adapting to a sustainable model of fashion can be challenging but it isn't impossible.

That's why action matters. Profiled by *Teen Vogue* and *The View*, twenty-one-year-old entrepreneur, philanthropist, three-time TED speaker, sustainability consultant, artist, global activist, coder, and author Maya Penn has inspired people all over the world. She started her sustainable fashion company, Maya's Ideas, when she was eight years old.

"The more I learned about sustainable fashion and the environment, the more ideas I had. I knew I wanted to work to support environmental justice and humanitarian efforts, so I founded my nonprofit, Maya's Ideas for the Planet, in 2011, when I was twelve years old." Projects include providing eco-friendly sanitary pads to girls in Haiti, Senegal, and Cameroon. "It's been amazing to see the impact in the US and around the world."

As a self-described "really nerdy kid," Maya has always been exceptionally artistic and had an affinity for animals and nature, which inspired her research on the fashion industry's environmental impact. She says, at first "a lot of my work included education because sustainable fashion wasn't mainstream back then. People didn't understand the impact that the industry has on water, toxic chemicals, greenhouse gas emissions, and labor."

Things are changing. Some companies report the carbon footprint of each product. Rental-wear companies are encouraging a new model of a shared fashion experience. These examples demonstrate the circular economy, defined as "a model of production and consumption, which involves sharing, leasing, reusing, repairing, refurbishing, and recycling existing materials and products as long as possible."[35]

We have an opportunity to change our behavior and send powerful market signals to the fashion world.

 In addition to choosing sustainable fashion, buy used items and repair your clothing. Swap clothes you don't wear anymore with friends—make a party of it! What other simple changes could make your closet more sustainable or reflect the concept "less is more"?

Take action by supporting the nonprofit Fashion Revolution to urge your favorite brands to protect people and the planet.

# Know that We Can Fix It & It's Not Too Late

We can fix the climate crisis. The technology to curb greenhouse gas emissions exists right now, and there are quick fixes that could make big impacts in the next decade. The world must mobilize quickly. But we need the political will to demand action. That happens when people like you and me stand up for change in our communities, state, and nation, and eventually, globally.

## Technology

The technology exists for more than *a 70 percent reduction* of current carbon emissions in the next ten years.[36] Renewable energy is becoming cheaper and more available, and in 2023, global investment in clean energy matched fossil fuels for the first time.[37] Electric cars and trucks, wind farms, solar arrays on houses, and energy-efficient appliances, operations, and systems are becoming more commonplace. "Nature-based solutions" like soil conservation, sustainable agriculture, and ecosystem restoration to store more carbon on the planet are taking off.

# Ready Solutions

The think tank Rewiring America concludes that modernizing and restructuring the electrical grid of the United States in the next fifteen years could rapidly decarbonize our economy and also create twenty-five million jobs.[38] Re-electrifying the country with five already proven technologies could get us to net zero by 2035. Net zero means reaching a point where greenhouse gas emissions equals the amount removed in the atmosphere, so there is no net increase of carbon emissions. Here's how:

- Bring more wind and solar power plants online.
- Transition to electric vehicles.
- Incentivize rooftop solar panels.
- Promote energy-efficient heat pumps instead of gas or oil furnaces.
- Invest in battery technology to store clean power.

With these same technologies, another report from University of California Berkeley researchers determined that 90 percent of the nation's power could be carbon-free by 2035 without passing costs on to consumers.[39]

The nonprofit climate think tank Project Drawdown's top climate solutions include adoption of clean energy at scale, reduction in global food waste, increased global equity for girls and girls' access to education, more plant-based diets, and reforestation.[40]

## Mobilization

Overall daily greenhouse gas emissions dropped 17 to 25 percent during the first months of the coronavirus stay-at-home orders. According to the International Energy Agency, overall global carbon emissions dropped 6 percent in 2020, the largest drop in the post–World War II economy.[41]

That initial pandemic experience showed that if we have the political will and can sustain our effort, we can create a clean energy future. Collective action through policy change is the ultimate solution for global warming. Consider the Inflation Reduction Act, which became law in 2022. It provides nearly $400 billion in climate solutions funding with more than $40 billion in consumer tax breaks and incentives to get clean energy and energy efficiency at home and switch to heat pumps, induction stoves, and electric or hybrid vehicles.[42] We are making progress.

 Learn more about how we can fix the climate crisis through Project Drawdown's Drawdown Roadmap video series.

## DAY 12
# Reduce Your Food Waste

**M**y husband and I joke that Google documents keep the romance alive. We have a shared document and calendar for events and for . . . wait for it . . . the weekly menu. Menu planning has saved us so many times. Planning for reuse of meals or ingredients can reduce food waste and saves money and time. This simple trick has also immensely increased the quality of our life.

Food waste generates 8 percent of global carbon emissions. The *Washington Post* stated that "if food waste were a country, it would be the world's third-largest emitter of CO2, after China and the United States." Community-wide and at-home composting can make an enormous difference. More than 80 percent of food waste comes from consumers at home or from grocery stores and restaurants. While we can design better dining environments, provide smaller plates, address portion sizes, practice restaurant site composting, and repack partially damaged foods, the majority of food waste occurs at home.[43] Even with a tight budget, you can create healthy and environmentally friendly meals.

 If you're not already menu planning, start now. Set a weekly reminder so you will carve out time in your schedule to plan. If you're incorporating menu planning into your weekly habits, how can you reduce food waste by using the ingredients for multiple dishes?

If you're interested in composting, here is a simple video on how to get started.

# DAY 13
# Advocate for Climate Justice

Equity is a necessary element to addressing the climate crisis. The concept of "climate justice" is rooted in the fact that people who contribute the least amount of carbon pollution are suffering the most. This is true racially, economically, and intergenerationally. Black, Indigenous, and People of Color and low-income communities are harder hit by climate change extremes. Study after study has shown that Black communities experience higher rates of air pollution and associated health effects like asthma and heart disease. High temperatures and air pollution are linked to pregnancy complications, and Black women are hurt at rates higher than any other group.[44] Heavy-polluting industrial sites are frequently located near communities of color. When Hurricane Katrina hit New Orleans in 2015, 30 percent of New Orleans residents didn't have cars, which meant it was harder for them to evacuate. And when stranded residents eventually relocated, many did so permanently because they couldn't return to New Orleans without access to transportation.[45]

Historically underserved communities often reside in areas with poor infrastructure and lack access to air conditioning, heating, and good insulation, making them more susceptible to weather extremes. It's also typically harder for Black, Latinx, Asian, and communities of color to gain access to fire or flood insurance to rebuild after a disaster or to pay for medical bills. Indigenous communities also suffer disparate impacts of toxic pollution, warming climates, and droughts. This inequity is true internationally as well.[46]

As we envision the future, let's clearly define how equity plays into this positive worldview for a just transition to a greener, healthier future and how we can work to protect the most vulnerable. A cultural shift for big climate solutions will take more than a mindset. We need a positive vision to galvanize support. If we can see it, feel it, and believe it, then we can achieve it.

 What is 'climate justice'? To learn more, read:

# Make the Switch to Clean Beauty Products

The scale of environmental issues can be mind-boggling, even paralyzing. "But we have to take action. We can't be complacent. We have to use our cumulative brain power and move forward together," remarked Gregg Renfrew, founder and former CEO of the beauty brand Beautycounter.

Beautycounter has been recognized as one of Fast Company's most innovative companies year after year. Gregg's leadership disrupted the beauty industry and made reforming the 1938 federal cosmetics law a household issue. In December 2022, ten years after she and her team began lobbying for cosmetics reform, the Modernization of Cosmetics Regulation Act of 2022 was signed into law.

"In 2006, I watched *An Inconvenient Truth* by Al Gore. It was a wake-up call. I was aware of global warming but realized I wasn't doing nearly enough and made the connection that what was detrimental to our earth was likely detrimental to our personal health. Gore's documentary helped inspire me to create Beautycounter. I knew from the start that my company would prioritize environmental health and sustainability."

With twenty years of experience working for prominent brands, Gregg launched Beautycounter and established the "never list" of toxic chemicals that the company wouldn't use in its product formulations. She then recruited women across the country to join her as consultants to tell the story of safer ingredients, to sell safer products in their communities, and to lobby Congress to fix our broken toxics laws.

Gregg sees extraordinary possibilities for businesses in climate solutions and in creating a regenerative economy. "Companies will have to recognize the challenges of the climate crisis, and many have. We can work toward a triple bottom line of people, profit, planet, because that's what consumers expect. Consumers want to be part of a community like that."

As she explained, "Gen Z knows the challenges they're facing with the climate crisis. They are bright, and they understand that the solutions are right in front of us. Earth can heal itself if we let it." One thing Gregg has learned is that "the small things matter. We have to take control of what we can control." Choosing to buy cosmetics without toxic ingredients and advocating for safe products is essential to environmental health.

Read and download The Never List™.

Using The Never List, review the labels of the beauty products that you use. How many of your personal care products include toxic ingredients?

# DAY 15
# Lean into the Law of Identity

We must embrace what I call the Law of Identity to make habits—such as a daily ritual of sustainability—stick. If you view the ritual as part of *who you are*, you're more likely to sustain it. For example, James Clear points out that if you want to be known as someone who is physically fit, you'd ask yourself, *What would a healthy person do every day to achieve their goals?* Then you'd take small, daily actions—such as adding more steps to your day, skipping your morning venti latte in favor of a mug of green tea, and so forth—and keep checking in until you've reached your goal. As Clear communicates about the power of identity, "Decide the type of person you want to be. Prove it to yourself with small wins."[47]

Everyone has a different skill set and personality to help lead the necessary culture change for big, positive action. Through this book, I want you to tap into what you likely already know about yourself and how you show up in service. Ask yourself who you want to be: How do you care for the important people in your life? Are you a fixer? A listener? The person who plans meals for someone who is sick? The person who sends a friend a good book or a poignant song when they are grieving?

When you're developing your daily practice of sustainability, remember that living an eco-friendly life isn't merely doing the actions but aligning the activities with your identity. As American politician, author, and educator Shirley Chisholm once said, "Service is the rent we pay for the privilege of living on this earth."[48] How do you want to pay it forward?

You don't have to figure it out right now but acknowledging your skill set, your unique strengths and interests, and how you can use them to live a greener life will set you up for success.

Later, we'll get into a more in-depth conversation about aligning your activities with your identity. For now, I want you to answer Shirley Chisholm's question. How do you want to pay it forward?

## DAY 16
# Link Air Pollution, Wildfires, & Climate Change

It's now commonplace to see wildfire-fueled orange skies serve as a stark backdrop to city skylines around the globe. In the summer of 2023, more than 120 million people in the United States were under threat from air pollution warnings—not just from industrial pollution but from pollution called "particulate matter" released from wildfires raging across Canada.[49] Public health officials advised wearing N-95 masks to protect citizens from "code red" unhealthy air. Air pollution triggers emergency room visits from heart disease, stroke, diabetes, asthma, poor lung function, and premature births.[50]

I live in Montana, where fire season is part of summer. But what we're now experiencing is not normal. Fire ecology professor Phil Higuera of the University of Montana recently explained that we're in uncharted territory when it comes to fires in the Northern Rockies. Based on sediment core samples, his team determined that we're in a catastrophic fire cycle in the Northwest, which hasn't happened in two thousand years. No matter the cause of these mega-fires in the Western United States—lightning strikes, arson, downed power lines, accidents—extended drought, extreme heat, and "wavy" jet streams from climate change will continue to make fire season more intense and more frequent in the future. Thoughtful forest management is necessary. Indigenous controlled-burning techniques, which were forbidden by the federal government for more than a century, could prove a powerful approach to forest management to mitigate these destructive, climate change–fueled fires.[51]

Extreme weather will disrupt the lives of everyone on the planet in ways most of us can't fathom. Scientists are encouraging us to adapt to climate change as we wait for countries to enact global solutions. Meanwhile, fossil fuel companies have spent billions of dollars on slick marketing and lobbying campaigns to cast doubt on whether the climate crisis is even real.[52] To counteract this narrative from Big Oil, we must be ready and willing to share our knowledge with family, friends, and our community and demand that our elected officials take action.

To protect yourself during "code red" air quality alerts, wear a N-95 mask, consider buying an air purifier with a HEPA filter, and don't go outdoors unless it's absolutely necessary. Advocate for 100 percent clean energy, switch your gas oven to an induction stove, and don't leave your car running or "idling" when you're not using it.

Learn more about the air quality in your community and check out the AirNow website here:

## DAY 17
# Break Free from Plastic

According to the World Economic Forum, by 2050 there will be more plastic than fish in our oceans. In addition to the deepest part of the ocean, plastic particles have been found in rainwater, Arctic ice cores, and snow in Antarctica.[53] More than 90 percent of all the plastic ever produced has wound up either being burned in incinerators, landfilled, exported, or discarded into streets, waterways, oceans, etc.[54] (So much for recycling!) Plastic pollutes our bodies too.

A study commissioned by the World Wildlife Fund found that we eat about a credit-card-size worth of plastic each week.[55] Yuck.

That's where groups like the Plastic Pollution Coalition (PPC) come in. "What solidified my understanding of the connection between health and the environment was my mom's passing from breast cancer when I was just thirteen years old and she was only forty-three."

Julia Cohen is a cofounder and the managing director of PPC, a global alliance of organizations, businesses, and thought leaders working toward a more just, equitable world free of plastic pollution and its toxic impact. She collaborated with her sister, artist Dianna Cohen, to create PPC in 2009.

"I had the policy and public health background; Dianna had the art and creative experience," Julia said. After a meeting at Google, Dianna discovered that there were many others focused on the global onslaught of plastic pollution.

"We wanted to create a big tent coalition where groups around the world could collaborate and gather to talk about and take action to stop plastic pollution." PPC now has more than 1,500 coalition members from seventy-five countries and global youth ambassadors fighting to end plastic pollution. You can make a difference by:

- Refusing single-use plastic whenever and wherever possible. Choose items that are not packaged in plastic, and carry your own reusable bags, containers, and utensils.
- Reuse durable, non-toxic straws, utensils, to-go containers, bottles, bags, and other items. Choose glass, paper, stainless steel, wood, ceramic, and bamboo over plastic.

- Reduce your plastic footprint by cutting down purchases that are packaged in plastic. If it will leave behind plastic pollution, don't buy it.
- Rethink the ways you can refuse, reduce, or reuse.

 Take action by signing a PPC petition to stop plastic pollution and its toxic impacts here:

## DAY 18
# Protect Our Public Lands & Know Their History

**M**anaged by the US Department of Interior, our national park system extends to all fifty states, the District of Columbia, and US territories, with 425 individual units consisting of more than eighty-five million total acres. The Bureau of Land Management (BLM) and the Agriculture Department's US Forest Service also manage millions of acres of inspiring landscapes on our behalf. The Congressional Research Service estimates that the federal government owns around 28 percent of the United States (640 million acres).[56] These public spaces have become an important part of the American experience, but they are also Native American communities' ancestral lands. Learning about the creation of each park or public land area is a powerful way to honor the land. Sometimes the story isn't pretty. More public lands are being managed jointly with tribes who once inhabited the areas now considered national parks or forests.[57]

And of course, global warming will change these landscapes significantly.

Take Yellowstone, for example. In 1872, President Grant signed into law legislation creating Yellowstone, the world's first national park, for the "benefit and enjoyment of the people." This magical area has ten thousand thermal features, more geysers than anywhere else in the world. It also contains Yellowstone Lake, the Lamar Valley, and Hayden Valley—some of the most spectacular scenery in the United States. Called America's Serengeti, it's known for its remarkable wildlife, as well as herds of bison, wolves, grizzly bears, elk, pronghorn, and sandhill cranes. Twenty-six Native American tribes considered the area that defines the park as sacred hunting grounds.

Like many national parks, this famous ecosystem is under threat from climate change.[58] Since 1948, Yellowstone temperatures have risen nearly two degrees. This warming means big changes are in store, including likely changing the plant composition in Yellowstone.

The growing season has expanded by more than thirty days, which will likely lead to more invasive plants like cheatgrass. The range of keystone species like the white bark pine, which serve as food sources for the Clark's nutcracker and grizzly bears, will become smaller. Scientists report changes in bison migration and fish spawning as waters rise and spring begins earlier. The shorter winter will likely extend wildfire season. There is little doubt that Yellowstone will still be here in another 150 years. But the question remains: What will it be like for our great-great-grandkids and the wildlife?

What's your favorite national park or public land? What do you know about its history? Do you know what tribes are associated with that land? How do you think climate change may impact this important place to you?

To find out about the Native American connections to the lands you love and live on, visit this resource here:

# Find Out Where Your Energy Comes From & Go Green

There are so many areas to learn about in the clean energy space. Investment in clean energy surpassed dirty oil and gas production for the first time in 2023. With the passage of the historic Inflation Reduction Act, look for tax incentives at the federal, state, and local level to switch to green energy at your home.

## Solar

Neighborhood solar gardens are a new concept in many states. The idea is that consumers "subscribe" to a solar installation owned by community members, including renters and those who can't afford solar installation.

For homeowners, installing solar panels continues to become easier and more affordable. The average solar panel cost for a 3,000-square-foot home ranges from $15,000 to $30,000—but the prices for the panels and installation are decreasing, especially as competition increases. If the initial cost is a barrier, consider rent-to-own solar panels through groups like Mosaic, EnergySage, and SunRun, which provide financing arrangements. In lower-income areas, the nonprofit Grid Alternatives is working to make clean solar energy affordable and accessible to ensure the clean energy revolution is grounded in equity.

## Switch to Clean Energy

According to the Earth Day Network, 600 out of 3,300 utilities across the country allow consumers to switch energy providers.[59] Depending on where you live, you may have the option of switching to 100 percent renewable energy. This usually means your energy is generated in the same location, but your payment is used to buy renewable energy certificates to support green energy in other locations. The other option is to buy these certificates yourself, which ensures that for every kilowatt used, you purchase one kilowatt-hour of clean energy.[60]

## Electrify Your Kitchen

Check out tax incentives and rebates to make the switch from a gas stove to an electric induction one. Recent reports show that gas stoves emit toxic air pollutants that can hurt our lungs. Thirteen percent of childhood asthma cases can be linked to gas stove use. In addition, gas creates carbon pollution and stoves often leak methane, a powerful greenhouse gas. Making the switch to an induction stove can reduce these emissions and promote cleaner energy. You may also be eligible for incentives to swap out your gas heat pump to an electric one.[61]

It's a lot to absorb because tackling the climate crisis is a big task, but we can break it down into smaller steps. Your daily habits will drive the movement by changing our society's mindset and fostering a culture that embraces a clean energy economy.

Take action on clean energy by:

Telling family members what you've learned about climate change and engaging with your community about climate solutions.

Calling your utility company and asking where your energy comes from and what percentage of clean energy they produce.

Lobbying your local, state, and federal legislators for comprehensive climate change policies and speedy implementation of the climate provisions of the Inflation Reduction Act.

Consider the unequal impacts of climate change. With record heat and extreme weather events, the most vulnerable are likely in danger, either from heat stroke without air

conditioning or going without heat in freezing weather. What
needs to change at the federal level? State level? In your
community?

# DAY 20
# Stand with & Encourage Young People

---

**M**y involvement in the environmental movement was incremental.
Ever since I was a kid, I knew we needed a clean earth. I wanted
to be on the right side of history preserving our natural resources,"
stated Dr. Shane Doyle, president of Native Nexus, an educational
consulting firm. As a member of the Apsáalooke nation from the
Crow Agency, Montana, Shane was glad to help when nonprofits
would reach out for a land acknowledgment or an honor song or
prayer. "There was an ebb and flow of people reaching out while I
was in graduate school. Then I became a spokesperson for several
campaigns about public lands. I didn't see myself as an environmental advocate. More and more I became engaged in climate change
and made sure that Native voices were heard."

Shane shared, "I am hopeful for our planet because we have
more knowledge than we've ever had. We know what we are up
against, and the solutions are there. We need to tell young people
that we can make a better day. Don't sit back and wait. Quietly go
about garnering support. And lead with love."

When asked about Generation Z, he pointed out that this new generation will make an impact on public policies and law and that they're committed to environmental issues. Two of Shane's daughters, for example, are plaintiffs in a youth climate action lawsuit, *Held v. Montana*, filed against the State of Montana for failing its constitutional duty to guarantee Gen Z the right to clean air and water. This historic case not only marked the first constitutional youth climate action lawsuit to go to trial but also the first winning verdict for youth plaintiffs. Shane also thinks we must remind young people of the positive social changes we've witnessed. "I've seen a sea of change in my lifetime that I would have never thought possible."

With respect to where to start in climate action, Shane emphasized Plains Indian culture: "They lived a ceremonial life for thousands of years. They were tuned into the natural world, our space, community, animals. If we walk like that and are aligned with an environmental ethic, we become role models for those around us."

He recommended being intentional about the small decisions we make every day, like how to travel or how much and what we eat. "You don't have to be religious to live life in a ceremonial way. It's a mindset and spirit of divine circumstance."

Have you talked with the young people you love about the future we're leaving them? If not, ask them how they feel about the climate crisis. Let them know they are not alone, and that it's not their sole responsibility to fix this. Tell them that you will support them in climate action.

Check out my TED talk, "Think Like An Awesome Ancestor: A Daily Practice to Ease Eco-Anxiety".

# DAY 21
# Choose Organic Food When You Can

D oes buying organic really matter? Yes, it absolutely does. In the United States, the USDA Organic seal means that food is grown without hormones, toxic pesticides, synthetic fertilizers, radiation, or genetically modified organisms (GMOs). Genetically modified organisms are defined as "containing DNA that has been altered using genetic engineering."[62] Most of the GMOs in the United States are plants, including 90 percent of corn, sugar beets, and soybeans.[63]

GMOs remain controversial. One of the biggest issues is use of the toxic weed killer glyphosate with genetically modified plants, also known as RoundUp Ready seeds. These plants are designed to be resistant to the herbicide. Glyphosate kills the weeds nearby by blocking out an enzyme that regulates plant growth, but the genetically modified plant doesn't die once it's sprayed. Unfortunately, the weeds around the genetically modified plants have also become resistant to glyphosate. Therefore, farmers spray more toxic weed killer; there's been a 500 percent increase since it was introduced on the US market.[64] The Interagency for Research on Cancer categorizes glyphosate as "a probable carcinogen" for humans. The EPA has declared that the chemical compound is *not* cancerous. On the other hand, the Agency for Toxic Substances and Disease Registry released a study showing exposure brought increased risk for non-Hodgkin's lymphoma and multiple myeloma.[65]

Avoid GMOs by buying organic when you can. Studies show that eating organic food can reduce the presence of pesticides in your body.[66] And because certified organic food is grown without toxic pesticides, it's better for the environment and your health. Generally, foods with thick skins like avocados, onions, and melons almost always show up as low pesticide residue options on the Environmental Working Group's (EWG's) Shopper's Guide to Pesticides in Produce, a great tool for choosing when to buy organic. Choose organic for leafy greens, apples, and strawberries. You can also consider shopping local or even trying out the 100-mile diet, only buying food grown within a 100-mile radius of where you live.

Since 2008, organic sales have tripled, but land used in organic production hasn't increased as fast. Organic farming reflects only 0.6 percent of the production acreage in the United States.[67] Barriers to converting farms to organic are largely financial, which is why providing support to farmers to make the transition to either organic or regenerative agriculture can have significant climate benefits.

Next time you're food shopping, choose organic when you can—especially for meat, berries, and leafy greens—to avoid GMOs. Look for the USDA organic seal.

Think about what you ate today. Where did it come from? How was it grown? How can you make better decisions for your health and the planet?

To find non-GMO products, check out the Non-GMO Project's brand identifier here:

## DAY 22
# Green Your Holidays & Gifting

When you're decorating for traditional holidays or birthdays, embrace these quick hacks so you create less waste and a more meaningful celebration.

1. **Decorate without single-use items.**
   - Decorating for celebrations is so much fun. If possible, try not to buy any new decorations. Use what you have. Make it a trip down memory lane. Consider a decoration swap or a party, or make your own.
   - Get crafty. Make Pinterest your best friend. Use twigs, leaves, and flowers for your decorations.
   - For garlands and tablescapes, use real plants or consider trimming them from your tree. Compost them afterward.
   - Recycle those paper boxes. Americans do a good job of this. We recycle about 90 percent of our boxes each year. If you have bubble wrap, save it and reuse it.
   - Consider washi tape instead of plastic tape.
   - Use butcher paper and decorate it instead of plastic wrapping paper, if possible. Decorate the paper or ask your kids to.
   - Reuse wrapping paper and gift bags.
   - Use twine to tie up presents.
   - Store your paper from years past.
2. **Embrace the motto "Experiences, not things!"**
   - Think about trips, excursions to a museum, lessons like voice or painting together, a hike, walking outside, a dinner out, seeing a show, going to the movies, a family outing to the library—something that doesn't involve "stuff."

- Think quality over quantity. What are a few things your family or friends really want? Make those a high-quality item, then don't worry about the number of gifts under the tree or at the holiday party.
- Consider resale, secondhand, or gently used items. And there is nothing wrong with regifting! Seriously!
- Ask your family what they want! You'd be surprised. They likely want to spend time with you. Teens and other family members may actually appreciate cash instead of stuff.

3. **Green your parties.**
   - Use real stuff. Bring out your fancy dishes or use your everyday wear.
   - Plan that kitchen cleanup. Make it an intergenerational group or ask a friend in advance to stay after the party to help you clean up.
   - Remember less is more.
   - Think about a plant-based party—with vegan dish options.
   - Focus on the company, not the stuff.
   - Make the gift a donation to a favorite charity instead of a physical item.
   - If you're hosting a fancy celebration, consider renting your outfit, borrowing a friend's, or shopping secondhand. Or only buy high-quality clothing that will last.

# Practice the Law of Amplification

In 2017, Robert Kelly, an international relations expert on the conflict between North and South Korea, readied himself for a live BBC interview from his home. Shortly after the cameras started rolling, Robert's four-year-old daughter, Marion, bounced into his office to see what was going on. Her adorable glasses and now-famous and unmistakable walk showed her exuberance about what her dad was doing. Then, with almost perfect comedic timing, a baby walker appeared in the doorway. Robert's infant son was ready to join the excitement. Moments later Robert's wife, Jung-a-Kim, darted in to pull the kids out of the office. As Robert smiled and chortled a "Pardon me" to the announcer, his wife ducked to avoid the camera angle and scurried the kids back through the doorway. As if scripted, she then athletically lunged to close the door behind her. To date, this forty-three-second video clip has fifty-five million views on YouTube.[68]

The joyful essence amplified the moment. It wasn't staged. It was real life, both vulnerable and funny.

When most of us think about the internet, we focus on the downsides. Of course, there are legitimate concerns about digital addiction, nature deficit disorder, and all the negativity and shame that abound online—especially during election season. This viral clip, however, shows the powerful force of joy that can be amplified online, even if manufactured algorithms are pushing for more extreme content and negative engagement.

Peer-reviewed psychological studies show that joy is contagious. This finding is the essence of the Law of Amplification, which means that positive actions lead to more positive actions, and that sharing joy inspires others. The Law of Amplification is important to movement building. A 2008 study by Harvard and the University of California showed that happy people amplify their happiness to spouses, neighbors, and friends. Someone who experiences happiness and shares it can increase a spouse's happiness by 8 percent and a next-door neighbor's by 34 percent. Even better, that positive exchange can extend to three degrees of friends, reaching outside of the happy person's network.[69]

Sharing makes a difference, and when you share your joyful experiences, the impact of the repeated daily action amplifies. Recruiting friends and family to try out these eco-friendly lifestyle changes creates an opportunity to fight the feeling of being overwhelmed and to take action for a greener world. As more join and our actions multiply, you can help the culture shift.[70]

Have you had a personal experience with the Law of Amplification? What happened and what was the outcome? For example, perhaps a friend shared a social media message or inspiring story that made you think differently or take action.

Who can you invite to join your eco-friendly lifestyle changes and help with the culture shift?

# Research Where Your Dollars Go

I got involved in environmental justice from my early days of listening and learning. My grandfather was involved with the union movement and civil rights in the coal mines of Appalachia," reminisced Mustafa Santiago Ali, a vice president of the National Wildlife Federation. "My dad focused on civil rights as well, so I was watching and thinking critically about solutions."

Mustafa is a regular expert in national print and television media, including MSNBC, CNN, and VICE. He currently cohosts a live radio show and podcast, *Think 100%: The Coolest Show on Climate Change*, with singer and actress Antonique Smith and civil rights icon Reverend Lennox Yearwood. His unyielding compassion energizes others to stand up for a greener, more equitable future.

As a child, Mustafa remembers seeing "good, hardworking folks exposed to toxic chemicals that were making them sick. I've also seen the beautiful nature around me destroyed by strip mining and its devastating impacts on community health in West Virginia. At sixteen, I knew this work was my calling."

Mustafa joined the Environmental Protection Agency (EPA) as a student and helped found the EPA's Office of Environmental Justice. He worked there for twenty-four years and says that there are "huge sets of challenges in front of us from the climate crisis" and "serious concerns with global warming, but I also see innovation and ingenuity. We can transform our society with new technologies to build a clean and just energy future. Things can be different. We can invest our time, bodies, and intellect—not to hope for change but to demand it."

In Mustafa's view, the most promising climate solution is environmental justice. "We can't win climate change if we don't address environmental racism," he said. "BIPOC communities have been hit hard by pollution. That's where fossil fuel refineries and toxic chemical plants are located; freeways, train tracks, and other transportation policies impact BIPOC towns too," he added. "In agriculture, decades of discrimination not only destroyed Black farming communities but meant we lost important lessons of regenerative agriculture." Mustafa underscored that "Black and brown folks are hurt the most from these environmental policies and practices. It's systemic."

His advice for those who want to be part of the climate movement is "first, get educated. Use trusted resources, research, and learn. Second, get engaged. There are so many frontline organizations that needs support. Finally, Know where your dollars go. Where does your electricity come from? Where do your clothes come from? Let's better understand who we're supporting. Are we paying people to shorten our lives? How do we better use our individual resources to support the future we want to experience?"

 Google "Cancer Alley" and learn more about environmental justice, environmental racism, and the impacts of the petrochemical industry on Black, Latinx, and Indigenous communities. Research what companies you are supporting directly or indirectly.

Use this tool from As You Sow to find out how to green your personal investments.

## DAY 25
# View Nature as a Climate Solution

Habitat conservation provides space for animals to roam, eat, and live, but it also provides carbon sinks with dense trees and protects against desertification. A carbon sink soaks up carbon dioxide from the atmosphere. This concept of using nature as a climate solution is called "climate resilience." The three most effective strategies are stopping deforestation, protecting soil, and limiting desertification.

Deforestation contributes 11 percent of all global greenhouse gas emissions caused by humans.[71] Demand for agriculture, mining, and housing leads to the burning of tropical forests, mostly in South America, to create room for grazing and development. The pressure on Indigenous people in these areas is significant. When trees are cut or burned down, the result is a triple whammy of carbon emissions because (1) the trees are no longer carbon sinks; (2) the trees release carbon when they're cleared by burning; and (3) the cut-down forests make room for livestock that create more carbon emissions.[72] Project Drawdown determined that curbing deforestation, especially in the tropics, is a worthy climate solution.

Lands in dry areas of the world are turning into deserts as trees and soil are stripped away by unsustainable farming practices resulting in soil erosion, overgrazing, and faulty irrigation practices. Called "desertification," this process means that there is overall decline in land productivity, and "over 100 countries and more than 1 billion people in the world are facing the threat of desertification."[73] When grassland or forests are converted to agriculture, carbon stored in the soil is released, which increases carbon emissions. That's why healthy soils and soil conservation are so important to address climate change. The Rodale Institute predicts that regenerative agriculture could sequester more than the United States' total annual carbon dioxide emissions.[74]

We can make a difference through nature-based solutions: planting trees, using sustainable agriculture methods, practicing soil conservation, and supporting coastal restoration.

"Blue carbon" refers to ocean and coastal vegetation that stores carbon.[75] Think salt marshes, mangroves, and sea grasses, which store carbon in coastal soil. Coastal reforestation and wetlands mitigation techniques can be powerful tools in climate solutions. In wetland areas, mangroves are particularly important as conservation strategies.

Learn more about nature-based climate solutions here:

Find out about ocean-based nature solutions from the report published by RARE.

## DAY 26
# Become More Aware & Process the Hard Truths

This story will likely stay with you for several days, and then you'll see why I told it. It's about love and compassion.

In September 2020, as raging wildfires roared across the American West, reporter Capi Lynn of the *Salem Stateman Journal* recounted the story of a family outside of Salem, Oregon, who received fire evacuation orders in the middle of the night. The husband had already left the house to borrow a friend's trailer to gather family belongings. When he returned, police had barricaded the road to his home because the fire had spread. He turned his car around to find a way through to the house, and he saw a woman on the side of the road. She was severely burned and looked like she was naked and barefoot. He pulled over, asked if he could help her, and told her he was looking for his wife. She said, "I am your wife."

Unrecognizable to her husband, the wife had walked three miles in fire and heat so intense that even her clothes and shoes melted. Earlier, when she realized that fire would soon engulf the house and her husband hadn't returned, she instructed their thirteen-year-old son to run for his life. In that instant she also made the heartbreaking decision to leave her mother, a seventy-one-year-old invalid, at home.

After the husband got his wife medical help, he and rescue teams searched for his son for days. Then they received the news from the authorities. Instead of running away, the son went back to the house and tried to save his grandma. He died in the driver's seat of the family car; his dog was draped over him. Authorities found his grandma's remains in the car with them.

This story is a climate change story.[76]

This is our future.

Despite the heartbreaking losses, this story also illustrates compassion, resilience, and community connection in action. According to GoFundMe, the wife has made a successful recovery. The community raised over $300,000 for the family.[77]

There are countless other examples of devastation from climate change and there will be more if we don't act on climate policy.

Most of us would rather act like everything's fine. Some experience shame when they fully realize what we're leaving future generations. For others, the climate crisis is so enormous it's hard to talk about.

That enormity leads to another problem: systemic desensitization, which means that when the numbers become so large, we can't appreciate the loss. Global losses from wildfires, mudslides, and floods are on a scale that's hard to process. That's why sharing stories like this one personalizes this crisis and inspires us to engage compassionately with our communities and each other.

Right now, there's a lot of animosity and negativity in our culture. It's up to us to break through that pain, highlight what unites us, and focus on compassion and community connection. It's time to work together to create a positive vision for the future. And what's a stronger connector than the planet we all share?

 Sit and reflect. How can you be more compassionate with your community and the people you interact with every day?

# DAY 27
## Understand POPS and EDCS

Fifty years after they were banned, chemicals like DDT (an insecticide used in agriculture) and PCBs (manufactured chemicals made to suppress fires) are still showing up in our bodies and the bodies of newborn babies. Those chemicals are persistent organic pollutants, also known as POPs, which bioaccumulate through the food chain and never break down.[78] Internationally, twelve POPs, known as the Dirty Dozen, are part of an international treaty because of their adverse effects on humans and the environment. These include pesticides, industrial chemicals, and chemical manufacturing byproducts.

In the past fifty years, childhood leukemia, brain cancer, and hormone-related cancers have exploded. A growing number of scientists, doctors, and pediatricians believe that this trend doesn't merely reflect better diagnostic methods, but that widespread, low-dose, chronic exposure to industrial chemical pollution contributes to these diseases.[79] Many of these toxic industrial chemicals mimic estrogen and are called endocrine disruptors.

The scientific community has known about the impacts of endocrine disruptors on human health for a long time. In the 1996 landmark book *Our Stolen Future*, the authors explained that these low-dose exposures to hormone-mimicking substances could impact obesity, fertility, intelligence, asthma, diabetes, endocrine-related cancers, and ADHD.[80] More than twenty-five years later, the peer-reviewed science demonstrates harmful effects to our hormone system from these synthetic EDCs at lower and lower levels of exposure.[81] Common EDCs include the synthetic estrogen and plastics hardener BPA (bisphenol A), chemicals used in food packaging like PFAS, organophosphate pesticides, flame retardants, and drinking water contaminants like perchlorate and the herbicide atrazine.[82]

Climate change will make toxic industrial pollution worse and expose us to more POPs and EDCs. Higher temperatures mean that toxic chemicals can vaporize and off-gas from polluted sites and plastics at a faster rate, or break down into toxic byproducts. Heat can also increase the redistribution of POPs that have settled in soils. Keeping POPs "locked in soil" is a similar strategy to creating carbon sinks. Given that many EDCs pollute drinking water, intense floods and hurricanes may increase exposure through additional sewage overflows and extreme rainfall events.[83] Firefighters spray flame retardants from airplanes to control wildfires, increasing the toxic load in water and soil.

 Are you familiar with POPs or endocrine disruptors? If not, are you surprised they're in consumer goods? If yes, what do you do to try to avoid exposure?

## DAY 28
# Plant a Climate Victory Garden

Robin Hill-Emmons started the nonprofit food-justice organization Sow Much Good after her brother, who struggled with mental health issues and homelessness, needed help. After becoming his legal guardian and finding him proper care, Robin was surprised that her brother became borderline diabetic, even while his mental health improved. Robin realized that the processed food the housing facility provided him was the culprit. To her shock, he'd received better nutrition when he was unhoused. Robin left her corporate job, tore up her manicured suburban yard, and planted a garden of fresh vegetables for his facility.

"I used social media to tell people about my garden. More than fifty people showed up and helped me harvest. I told them that I wanted to feed people in a transitional housing group, and one of those people was my brother," Robin explained. "When neighbors would ask why there were so many cars parked on our street, I asked them to join us. And they did."

Sow Much Good expanded quickly, from supporting Robin's brother's facility to targeting the BIPOC community in low-wealth areas, creating a community-supported agriculture program, developing a farmers market, offering cooking classes, and sustainably growing more than twenty-six thousand pounds of food on more than nine acres in and around Charlotte, North Carolina. Robin's advocacy garnered significant media attention, including features in *People*, CNN, PBS, *Modern Farmer*, and *Southern Living*. Robin also received the CNN Hero Award for her leadership.

One of her key lessons in advocacy is that "the heart can't feel what the eyes can't see." She understands that homelessness, food justice, and climate change are interrelated on many levels, including the need for people to "see" the challenges.

In 2018, after eleven years of service, Robin wound down the nonprofit. By then other food justice groups had emerged to continue the work. "Through this experience, I knew that I could be a leader of others to create big solutions." Robin always fights for the underdog with her "finely calibrated moral compass." Systemic racism and the challenges of growing up in a low-wealth community motivated her to be a bright light for positive change.

Growing your own food can minimize waste and save money. During World War II, so-called victory gardens were a call to action for patriotism and to reduce the pressure on food suppliers during the war effort. If you don't have the resources to garden, consider planting some herbs on your windowsill or shopping at your local farmers market to get to know who grows your food.

Here is an insightful article from Columbia Climate School.

To learn more about food justice, check out the work of Soul Fire Farm.

## DAY 29
# Choose Green Cleaners

Getting smart about the cleaners you use in your home can make it safer, particularly if you have young children. The Cleveland Clinic has a handy outline of key categories of cleaners to watch out for,[84] including:

- **Oven cleaners**, which usually contain toxic chemicals that may be fatal if ingested. These chemicals can trigger asthma attacks too.
- **Avoid fragrance** in general because it includes phthalates, which are hormone disruptors and can cause allergies.
- **Avoid "quaternary ammonium compounds,"** which are used to kill germs in fabric softeners and air fresheners. These ingredients can trigger asthma.
- **Laundry detergents** that have cationic, anionic, or non-ionic enzymes on the label can cause nausea or vomiting if ingested.
- **Avoid dryer sheets**, which are usually loaded with perfume and can contain the carcinogens benzene and acetaldehyde.[85]

 Reminder that just because it has "green" on the label, doesn't mean it is. Check out the EWG Guide to Healthy Cleaning to verify.

Here are some easy do-it-yourself cleaner recipes from Good Housekeeping.

## DAY 30
# Create an Eco-Action Plan

The purpose of the Eco-Action Plan is to create a road map for your daily ritual of sustainability. Tailor it to meet your needs. Then share your progress with your circle. Here are the five steps:

### Step 1: Planning Time

What's important to you? What do you want to feel about nature, the earth, and the future? What green thing aligns with your interests?

### Step 2: Share and Listen

If you're going to engage others with your plan, share and take notes. Think about how your interests may differ and where you have common ground.

Our family came together to create an Eco-Action Plan. My older daughter Cady's biggest concern is energy policy. To tackle the root cause of climate change—dirty energy production—she knows we must create fundamental policy changes to shift to renewables. Susan is passionate about educating girls and providing them access to health care and family planning, to promote equity and increase innovation. Educating girls globally is one of the top ten solutions to climate change, according to Project Drawdown. My husband's work reflects his top priority, which is protecting wildlife habitats and public lands. My biggest concern is to ensure that elected officials act on climate. It was fun to talk through how we could apply our different gifts, work toward solutions, and chip away at the feeling of being overwhelmed.

## Step 3: Dive In and Learn

Research the issues you care about and set up a plan for action. Learning about the issues you care about can help guide your daily practice.

After our family's initial discussion, Cady started reading about energy policy, and Susan elected to learn more about girls' access to education. My husband decided to double down on his professional work in wildlife management. Our discussion inspired me to study the impacts of a plant-based diet on global warming and review the climate platforms of elected officials.

## Step 4: Create a Plan and Track Progress

Create your Eco-Action Plan and write down your priorities and action steps. Research shows that writing down your goals increases the likelihood of success. Recent studies confirm that "performative environmentalism," or individual action, creates the cultural momentum for policy reform.[86]

## Step 5: Celebrate and Look Ahead

Check in after your first month—that's Day 60 of this reader. What worked? What didn't? What's next? What action do you think had the most impact? What made you personally feel better?

# DAY 31
# Embrace Cathedral Thinking

I'm inspired by author Roman Krznaric, who writes about "cathedral thinking" and urges us to adopt an ethic of "longtermism" as we think of ourselves as ancestors.[87] Our days quickly sift through the hourglass, and therefore we should intentionally think about what we're building for the next generation. He points out that humans can excel at long-term thinking and highlights examples, including the Trans-Siberian Railway, the global eradication of smallpox, the Green Belt Movement in Africa, the US Constitution, and Yellowstone National Park.

Past generations thought about us—our needs for transportation, health care, and even wild places. Of course, they missed the mark a lot too, but these incredible advances benefit us all. In climate action, we must adopt this cathedral thinking and start acting in future generations' best interests.

Taking time to consider our personal and generational legacy is a worthwhile exercise. Thinking about life after you leave this planet is what Krznaric calls the "death nudge," a reminder of your own mortality. I prefer to call this phenomenon the "hourglass," not only as an homage to *Days of Our Lives* but also as a way to express the notion of coming to terms with the inevitable passage of time. Behavioral psychology research shows that as people become more aware of their mortality, they're more likely to act in future generations' interests. One study showed that participants who wrote a brief essay about their legacy and how they wanted to be remembered donated 45 percent more to charity than those who didn't write the essay.[88] Another study examined lawyers who first told clients that most people leave money to charity in their wills and then asked the clients if they'd like to donate after they died. This resulted in a 17 percent giving rate, nearly three times the average.[89] This research means that talking about your legacy, thinking about what cathedrals you'll help build, and engaging with the next generation can inspire change. Focusing on the hourglass can result in long-term thinking.

What do you want the next generation to know about you and your experience on this planet? How do you want to be remembered? What do you want your legacy to be?

Take the "Be an Awesome Ancestor" Pledge here:

# DAY 32
# Support Land Conservation

L and conservation is the long-term protection and management of unused or underused land resources. Conserving land can provide habitat for native plants and animals and enhance what's called "ecosystem services."

"I got into advocacy not because of the environment, but because of land," explained Dan Puskar, president and CEO of the nonprofit Public Lands Alliance. "I grew up on a three-square-block residential development in rural New Jersey, surrounded by forests and fields and creeks to play in. Every summer my family and I went to the lake district in Maine, but I didn't think about it as 'environmental' stuff."

During his junior year in college, he discovered an exciting opportunity to study abroad in Botswana, but the condition was that he had to take ecology. "I was learning from native peoples living near the Okavango Delta and exploring conservation, anthropology, and ecology." This experience changed his life. "I understood the profound relationship between land, place, people, and biodiversity. The scale of it opened my eyes."

In his role at the Public Lands Alliance, Dan follows the ins and outs of Capitol Hill and tracks how appropriations and proposed federal and state policies will affect local, state, and federal lands and the nonprofits that bolster them. "Public Lands Alliance exists to convene grassroots nonprofits, to share ideas on what works and what doesn't, and to serve as an engine for capacity-building. We also work on big issues in public lands today, like creating safe, inclusive spaces for all people to enjoy the outdoors."

Nonprofit partners to public lands can break down barriers and form opportunities for trust and inclusivity. "I was very lucky early in my public lands career to find a Sierra Club Colorado–organized backpacking trip for the LGBTQIA+ community to the San Juan mountains. It meant something to be able to backpack in a national forest with people like me," Dan recalled. More nonprofit partners are also advocating for more inclusive experiences for BIPOC and other under-resourced communities.

Dan remarked that "setting aside land for carbon sequestration and supporting soil health is an incredible opportunity to curb carbon emissions. The good news is that a lot of people care about land conservation. We don't need to have the same underlying motivations in order to work together. And there is no action too small. Talk to people who are doing things you're interested in. Local impact matters just as much as international impact. We need all of it."

 To learn more about land conservation, check out the Public Land Alliance and the Trust for Public Land.

## DAY 33
# Get to Know Your Watershed

How much do you think about the water in your life? Where does it come from? What's in it? When I started studying environmental law, I was stunned by how much I took for granted. In the United States our drinking water is considered some of the safest in the world. Yet hundreds of communities in this country don't have access to clean water.

Our bodies are 60 percent water, and water covers 71 percent of the planet. Only 3 percent is freshwater, and 1 percent is drinkable.

The simple definition of a watershed is "an area of land that channels rainfall and snowmelt to creeks, streams, and rivers, and eventually to outflow points such as reservoirs, bays, and the ocean."[90] Visualizing your community as a watershed is a compelling way to think about it as you become mindful about the water in your life.

Global warming endangers our water supply. Extreme rainfall events create standing water that accelerates dangerous waterborne illnesses. Flooding jeopardizes access to clean water as waste ponds and water treatment centers become inundated and sewage overflows into lakes, streams, and rivers.

## Sewage Overflows

When I worked on Capitol Hill, the senator's office would get tons of calls when rapid-rain events caused water treatment facilities to overflow in Milwaukee, resulting in sewage overflows into Lake Michigan. If the untreated wastewater isn't released into natural bodies of water, it can cause sewage backups in people's homes. These sewage releases mean automatic beach closures.

However, green design and infrastructure like greenways, riparian buffers, and rain gardens are dynamic tools to manage intense rainfall and stormwater runoff. Across the nation outdated water utility infrastructure needs significant investment for upgrades to protect public health. Creating more green spaces can be part of this effort.

## Toxic Algal Blooms

Warm water currents, low salinity, agricultural runoff, and fertilizers laden with phosphorus, nitrogen, and carbon all feed algae and generate overgrowth. When these tiny algae are out of control, they emit gases that contaminate the air and deplete oxygen in the water, making it toxic for wildlife and humans. Agricultural nitrate pollution demonstrates how the way we grow our food is connected to the planet's well-being and economic health. Communities who've experienced toxic algal outbreaks have spent more than $1 billion since 2010 on prevention and clean up.[91]

 Do you know about the water in your community? What watershed are you in? Where does your water come from? Take five minutes and google it. Write down what you learn.

## DAY 34
# Realize that Consumer Pressure Matters

A well-known example of how consumer action can affect manufacturing and policy is bisphenol-A, or BPA. This synthetic estrogen was commonly used as an epoxy resin to line canned foods, a strengthener for polycarbonate plastic, and a coating for paper receipts. But BPA in cans leached into the food itself. More than two hundred studies linked BPA to hormone disruption, and research concluded exposure correlated to everything from cancer to reproductive problems, obesity, early puberty, and heart disease.[92] Bowing to intense consumer pressure and lobbying by environmental health advocacy groups, many companies abandoned BPA. However, some companies still use the chemical and others switched to problematic substitutes that also act as endocrine disruptors.[93] We need the EPA to review classes of chemicals, not just individual chemicals.

Phasing out flame retardants represented another win—at first. The three big culprits were PBDEs (brominated flame retardants), Firemaster 500 (containing a chemical similar in structure to a phthalate, which has been banned in children's products because of cancer concerns), and chlorinated tris (the chemical banned from treating kids' pajamas in the 1970s). A 1975 California fire safety rule resulted in widespread application of chemical treatments to prevent fires. Exposures to these "flame retardant chemicals have been linked to everything from hyperactivity/attention deficit disorder, thyroid disruption and even breast cancer."[94]

Scientific research, advocacy, and consumer action led to passage of a California law in 2014 to change the fire safety rule and promote safer consumer goods. This win shows we can make progress, but the problem of regrettable substitution still exists.

These two examples demonstrate why policy reform, advocacy, and consumer pressure must dovetail. They're also a reminder that the work is never done.

Learn more about BPA and the problematic replacements for BPA.

## DAY 35

# Check & Replace Your Personal Care Products

Lead in lipstick? Seriously? How is that possible? In 2011, the FDA tested four hundred different lipsticks and lip glosses and found lead in every sample. Although the average contamination was only 1.11 parts per million, there were 7.19 parts per million in a popular drugstore brand. The FDA said the lead levels weren't a safety concern, but the general consensus of public health experts says no level of lead exposure is safe.[95]

Switching to safer, less toxic products can reduce the presence of EDCs in your body. For example, a 2016 study examined the body burden of teen girls. When they switched to products with fewer toxic chemicals, there was a significant drop of these endocrine disruptors—specifically in their blood.[96]

Why do these studies matter? Some scientists think there's

a connection between personal care products and early onset of puberty. Subtle hormone shifts in the body at levels as low as one part per billion or one part per trillion guide the complex onset of puberty. Research shows that early onset of puberty can lead to increased risk of depression, unwanted sexual attention, and the risk of dropping out of school.[97]

Key ingredients to avoid in personal care products include the following: parabens, phthalates, formaldehyde-releasing preservatives like DMDM hydantoin, skin lighteners, and diethanolamine (DEA). For sunscreens, choose titanium dioxide and zinc oxide, which provide strong protection and are the only two chemicals that the FDA declares as "generally recognized as safe and effective."[98] The term "reef-safe" isn't defined, so choose a sunscreen without oxybenzone (also an EDC) or octinoxate, which have been found to damage coral reefs.[99]

In December 2022, the first update to the federal cosmetic law since 1938 passed Congress. Called the "Modernization of Cosmetics Regulation Act of 2022," the new law provides FDA recall authority for faulty products, fragrance allergen labeling requirements, and maintenance of records of safety for ingredients. This is good news, but more needs to be done to protect consumers and promote safer products.

Companies continue to step up to show clean beauty is possible. Brands like Beautycounter, Tata Harper, W3LL People, and True Botanicals also offer clean beauty choices. Check out Black-owned clean beauty brands like the Laws of Nature Cosmetics, Luv & Co, and Rooted Women. But no matter the hype, check the labels. Unlike the "organic" label for food, there's no formal definition of "clean beauty" or "green beauty." The Made Safe and EWG Verified certification programs take much of the guesswork out of shopping.

 To stay updated on the latest news in cosmetics safety, you can follow organizations like Breast Cancer Prevention Partners, Safer Chemicals, Healthy Families, EWG, Made Safe, and the Campaign for Safe Cosmetics. Use the Skin Deep or Think Dirty apps to find better brands at the grocery store.

## DAY 36
# Think about Spirituality & a Greener Life

I t's a real battle we're in. Never forget that we're on the winning team. It might be tough, and we'll encounter valleys and climb high mountains, but we will prevail," insisted Reverend Dr. Durley. Recently recognized on the National Park Service's "International Civil Rights Walk of Fame" for his leadership, Reverend Dr. Durley is a civil rights icon. He walked with Dr. Martin Luther King Jr. in the 1963 March on Washington. The author of *I Am Amazed!*, Reverend Durley refers to himself as a *rewired* pastor, not retired one, given that his focus has shifted from a specific congregation to the planet, and he now serves as board chair of Interfaith Power & Light.

Reverend Dr. Durley's involvement in climate action started when actress Jane Fonda introduced him to her daughter-in-law, Laura Turner Seydel, to talk about climate change.

"At first, I wasn't convinced. I told them, 'You lost me at polar bears,'" he said. As he learned more, he became intrigued. "I saw the connections between public health, redlining, and climate justice. We were not just in a legislative battle, but an ethical and moral one."

Reverend Dr. Durley continued: "Once people understand that climate change impacts them personally the opportunities to get involved become natural. You can start by pulling tires out of the river like in a Chattahoochee River Cleanup day. Once you get involved, you see you can do even more."

Reverend Dr. Durley's advice continues: "We are in a clim-a-demic. There is no vaccine. Prevention is the cure. We have to keep warming to 1.5 degrees Celsius. The climate movement can't be selfish. It must be for the good of all. God created a balanced world. As a nation, are Americans ready to change our lifestyles? Even a little bit?"

 Consider your spirituality, faith, or code of ethics. What does living a greener life mean to you? How would you make the moral and ethical case for climate action?

Learn more about spirituality and climate action through Interfaith Power & Light.

## DAY 37

# Understand that Wildlife Conservation Works

It was a beautiful August day in the Hayden Valley of Yellowstone National Park. Through my binoculars, I watched a bison herd cross the Yellowstone River, first one by one, and then about twenty bison crossed the deep water. The slanting sun baked the landscape in a golden light. Seeing these enormous creatures in the fast-moving river astounded me, not because they were elegant as they swam, but because I realized that I was witnessing a conservation miracle.

As it colonized the West, the US government purposely killed American bison to destroy Native American cultural, economic, and food systems.[100] When Europeans first settled in the United States, 30 to 65 million American bison roamed North America.[101] By 1902, there were less than 200 wild bison left in the United States. That same year managers of Yellowstone National Park purchased twenty-one bison from private landowners and began a conservation restoration project. In 1907, the park managers established the now-famous Lamar Buffalo Ranch for bison recovery. Today Yellowstone National Park has roughly 4,500 bison within the park. Domestically, the US Fish and Wildlife Service estimates that more than 200,000 bison are in private herds.[102] In the early 1990s, the US government partnered with Native American tribes to transfer recovered bison to tribal lands. In 2016, Congress declared the bison the national mammal and proclaimed the first Saturday of November National Bison Day.

While recovery efforts are still controversial, this powerful conservation success story shows that conservation strategies work. We can make up for lost time. We have the science and technology. But we need the cultural and political will to protect our planet, wildlife, and ourselves.

Wildlife conservation protects human health and safeguards against climate change by teaching us compassion, protecting genetic diversity, and serving as carbon sinks as a nature-based solution to global warming. Additionally, connecting to nature is critical to our overall well-being.

 Think about what animals mean to you. Do you or did you have a special pet or a connection with an animal in the wild that helped bring you comfort, shaped who you are, or enabled you to enjoy the awe of nature?

To take action and support wildlife conservation, check out:

# DAY 38
# Examine What's in Your Pantry

C hemicals in food additives have a long history in the commercialization of food. You need a dictionary to figure out how to pronounce some of the chemical names. In the early 1900s milk manufacturers used chalk and plaster dust to make milk look whiter. Pesticides were added to canned vegetables to make them greener, and formaldehyde was used to preserve meat. As chronicled in *The Poison Squad*, USDA chemist Harvey Wiley pushed for food labeling after years of research and advocacy. Wiley enlisted twelve young male volunteers, whom the press deemed the "Poison Squad," to experiment with whether eating food with certain preservatives would damage their health. When the men invariably got sick, Wiley used the study to convince Congress to pass the Pure Food Act in 1906.[103] The point is that Congress took action only after scientific studies led to public outrage.

The Food and Drug Administration (FDA) regulates food additives. But out of the ten thousand chemical additives in food, only a small percentage have been publicly studied, and many companies don't have to disclose what's in their food products. Our food safety law allows a huge loophole, called "Generally Recognized As Safe," where companies can "self-certify" new chemicals as safe and put them in their products without telling the FDA.[104] About one thousand chemicals are in food without the company even notifying the FDA, which means we just have to trust the food companies to "make good choices."[105]

So how does this relate to climate change? Sugar, wheat, meat, and dairy—all energy-intensive products—form much of our ultra-processed food products, which can contain largely unstudied additives and preservatives. Learn more by following groups like NRDC, Consumer Reports, EWG, and Center for Science in the Public Interest (CSPI) that advocate for food additive reform.

What's my advice? Eat real food, buy organic when you can, and read food labels.

 Look at the food labels in your kitchen. Research any and every ingredient you don't know or can't pronounce.

Check out this article from Consumer Reports on state legislation that will ban five food additives of concern.

## DAY 39
# Pack Your "Go Kit"

**T**ake the Scout's code to heart. Be prepared. You never know when extreme weather may impact you and the people you love. The news headlines are filled with extreme weather events from wildfires to floods. Create a simple go kit in the event that you need to leave your home quickly. Make sure you have water, food, a phone charger, a battery, a change of clothes, extra medicine, and essential documents. We often forget phone numbers these days since they are programmed into our phones. Consider writing down the phone numbers of your important contacts, including your health insurance and homeowners or renter's insurance. Some companies now sell evacuation kits, ready-made "to go" bags with essentials. No matter what, being prepared can help ease anxiety about uncertain weather or emergencies. Besides, it's what the Red Cross has advised us for nearly a century.

To be prepared, check out the OneGreenThing "Go Kit" checklist.

# DAY 40
# Commit to Climate Conversations

**B** estselling author Dr. Mark Hyman runs the UltraWellness Center in Lenox, Massachusetts, serves as the Head of Strategy and Innovation for the Cleveland Clinic Center for Functional Medicine, and hosts one of the leading health podcasts, *The Doctor's Farmacy*. He's written more than fourteen *New York Times* bestselling books; is an expert on food, health, and the environment; and is a regular contributor to *The View*, *CBS This Morning*, and *Good Morning America*.

As a doctor who practices functional medicine, Mark credits his undergraduate degree in Buddhism from Cornell and his spirituality as his wake-up call for a life in service. "Bodhisattva is not about enlightenment for your own sake, but to help others and relieve suffering. It takes a worldview."

After college, Mark went on a backpacking trip and decided that serving others through healing was his calling. During medical school, functional medicine was referred to as "clinical ecology," which Mark points out is an appropriate name for how human health and the planet are intertwined. His unique educational background, combined with an ability to see the whole person, primed him for a future in functional medicine practice. Mark's work reaches millions of people through his medical practice, books, engaging online platform, popular podcasts, and television appearances.

In his current practice, some of Mark's patients are concerned about climate change and the horrible headlines. He tells them, "Either you can embrace hedonism and do nothing or decide to become an activist. Action helps alleviate some anxiety. How you want to contribute becomes the biggest question."

Mark sees a positive shift in the future as more people "embrace a holistic future and realize there's a lot you can do, individually and politically, to make a difference." He gave this advice on how to get involved in the climate movement: "Take action. Whether it's composting, voting, volunteering, donating, or trying out brands. These small steps make a difference for your personal mental and physical health but also your overall well-being. Empower others with knowledge, support, and encouragement."

When it comes to intergenerational partnerships, Mark recommends talking to your kids as the crucial first step. "As simple as it seems, making service and caring for the planet a conversation around the dinner table helps them become family values. Those conversations are a powerful force to influence others and shift the culture toward global climate solutions," said Mark. "Community is essential to fuel lasting culture change."

When talking at the dinner table you should acknowledge your own emotions, make sure the conversation is age-appropriate so all ages can comprehend, and pledge to learn together.

 If you need help with climate conversation starters, here are some to consider:

What do you know about climate change and what questions do you have?

What climate stories or experiences affect you the most?

As a household, how can we use less plastic, buy less stuff, or use less water?

Is anyone interested in creating a family garden? What should we grow? How will this impact our environment?

How do you think we can bring more people into the climate movement and encourage elected leaders to take action?

## DAY 41
# Race Against the Sixth Extinction

On average, eighty-two species a day go extinct.[106] The "Sixth Extinction" refers to the unsettling finding that we're likely in a phase similar to the other five mass extinctions the earth experienced over the past half-billion years. We can't necessarily stop it, but that doesn't mean we shouldn't try. Biodiversity is important because it protects human health, promotes planetary resilience, provides humans food and medicine, and fosters strong mental health. Around one million animal species and eight million plant species are at risk of going extinct "within decades."[107] Pollinators are at high risk, which is worrisome given that 75 percent of global food crops need bees and other insects to pollinate them.

However, there's still hope. In the United States, the 1960s and 1970s brought about a complete environmental awakening. In 1962, Rachel Carson published *Silent Spring*, which documented the impacts of industrial and chemical agriculture on our waterways and wildlife. She alerted the public to the impacts of the pesticide DDT on transforming bald eagles' eggshells into brittle linings, thereby killing their babies by breaking during incubation or failing to hatch. The agricultural industry full-on attacked her, but she persisted.[108]

Rachel Carson's work created momentum for the EPA, which President Richard Nixon created in 1970, with strong bipartisan support for environmental action. Then Congress passed a slew of environmental laws from 1970 to 1980, including the Clean Air Act amendments, the phaseout of lead in gasoline, regulations on pesticides and insecticides, clean water protections, and toxic site cleanup and liability laws.

The Endangered Species Act of 1973 fundamentally changed how our laws valued our relationship to wildlife and aims to prevent the extinction of listed species, which are declared as either "threatened" or "endangered." The law also protects "critical habitat" and prohibits what's called a "taking," which means to hunt, kill, harass, or otherwise jeopardize the species. Of the species listed, 99 percent remain on the planet today, and 68 percent are improving.[109] As these species recover, many other issues arise, including creating wildlife corridors for them to move freely as they migrate.

How have you seen environmental progress in your life (e.g., cleaner air, water, or energy or a new park)?

Look up endangered species in your state here:

## DAY 42
# Eat More Plants, Less Meat

Trying a plant-based diet is one of the easiest ways to help the environment. Animal agriculture accounts for 20 percent of global carbon emissions. Project Drawdown calculates that if 50 percent of the world's population adopted a plant-rich diet, we could save 65 gigatons of carbon emissions globally.[110] Of course, a plant-rich diet is healthier too.

Campaigns like Meatless Mondays, where you skip meat at least once a week, are popular. Close to 70 percent of Americans are experimenting with more plant-based protein.[111] According to the Natural Resources Defense Council (NRDC), the ten most climate-damaging foods, in order from most to least harmful based on emissions required to produce them, are beef, lamb, butter, shellfish, cheese, asparagus, pork, veal, chicken, and turkey.[112] In addition to the environmental impact of animal agriculture, many are fed antibiotics to fatten them up and bring them to slaughter faster. This overuse can contribute to antibiotic resistance. The bottom line is to choose organic meat and support regenerative farms when possible.

Again, the goal isn't perfection—it's a daily practice of taking an action to make change. Start small and see what happens.

 Commit to incorporating Meatless Mondays into your meal plan. If you're already a vegetarian or vegan, share your favorite plant-based recipes with friends. Consider hosting a plant-based potluck at your next get-together.

## DAY 43
# Sign Up for an Online Workshop

Attending workshops is a way you can build community no matter where you live. While you're doing your one green thing each day, it is important to stay connected to others because you can't go it alone. Compassion, community, and connection help ease eco-anxiety and create a sense of hope through action.

Workshops can introduce a new concept, new research, and new efforts that you may want to investigate further on your own. You'll also connect with others who can demonstrate and encourage the practices of a daily ritual of greener living. Ultimately, it is a safe space for you to share your experiences and ask questions.

 Scan the QR code below to learn about upcoming workshops.

## DAY 44

# Appreciate the Ground Under Your Feet

I was such an unlikely crusader for cleaning up the food supply," said Robyn O'Brien, known as "The Erin Brockovich of food." Robyn's surprising journey began when her youngest child was diagnosed with an egg allergy. This prompted her to dig into food allergies, including chemicals in food, the lack of transparency and regulation, and how food is grown. She wondered, *Are we allergic to food or to what's been done to it?*

In 2009 her findings became the bestselling book *The Unhealthy Truth: One Mother's Shocking Investigation into the Dangers of*

*America's Food Supply—and What Every Family Can Do to Protect Itself.* Her book sent shockwaves through the food industry. "I learned that when you get that type of pushback, you have hit a talking point that [the industry] cannot address," Robyn remarked.

Despite the food industry's pushback—or maybe because of it—word got out about Robyn's work. As the US public became aware that multinational companies were selling processed food with less harmful ingredients in other countries, the market shifted.

"The early part of my work was advising multinational companies that consumer demand for organic, real food wasn't a fad. It wasn't just moms being loud. What they were seeing was a fundamental change in how the twenty-first-century consumer was shopping, and that would disrupt their business model."

Eventually climate change—specifically soil health—emerged as Robyn's central focus. Her company, Sirona Ventures, finances farmers' transitions to regenerative and organic agriculture, which provides soil conservation carbon offsets for food company supply chains. "Conventional agriculture has practically killed the soil." Through regenerative agriculture, "we can return soil into the living sponge it's supposed to be that soaks up carbon and holds water." Her message to companies is this: "As you heal the soil, you're taking care of consumer health, farmer health, and climate health."

Managing soil health allows producers to work with the land versus against it. Having healthy soil reduces erosion, maximizes water infiltration, improves nutrient cycling, saves money on inputs, and improves the land's resiliency. When growing crops, we want soil to hold water and nutrients like a sponge, suppress pests and weeds that attack plants, sequester carbon from the atmosphere, and clean the water that flows through rivers, lakes, and aquifers.

 Listen to this radio interview on how regenerative agriculture works.

Follow the National Sustainable Agriculture Coalition here:

## DAY 45
# Learn What's in Your Tap Water

The Safe Drinking Water Act regulates tap water in the United States by declaring legal limits for water contaminants. It establishes a standard to protect human health, called a "public health goal." Instead of meeting the public health goal, the law mandates a balancing of the economic costs of compliance to clean up the water pollutant. Therefore, just because it meets the legal limit doesn't mean it's safe.[113]

The EPA sets regulations for more than ninety different contaminants in public drinking water, including *E. coli*, *Salmonella*, and *Cryptosporidium*. Water utilities are required to test their water for these contaminants and disclose results to the public through Consumer Confidence reports. The rest are considered "unregulated contaminants," including forever chemicals and more than 160 other contaminants.[114] EPA hasn't officially regulated a new contaminant under the Safe Drinking Water Act in twenty-five years and only recently recommended that "forever chemicals" be regulated. Some standards haven't been updated for fifty years. Scientific evidence has emerged that multiple unregulated contaminants in drinking water could cause harm at extremely low doses.

It's time for reform. The EPA should regulate water pollutants based on what's safe for public health and consider vulnerable populations like the elderly, pregnant women, and newborns. The law should regulate whole classes of chemicals, not one chemical at a time.[115]

The Clean Water Act, on the other hand, regulates pollution at the source from factory pipes and from nonpoint sources like agricultural runoff. States and the federal government need more financial resources for water infrastructure and for oversight of nonpoint source pollution, which contributes to significant water quality issues.

Climate change will create more pollution, extreme weather, and pressures on our water resources. Consider the following actions:

- **Get to know your water utility.** Attend local meetings and learn about the issues in your community. Support funding for infrastructure upgrades.
- **Conserve water.** Take shorter showers. Buy water-efficient appliances. Use a water-efficient showerhead and toilet. Rethink your landscaping.
- **Know the top water contaminants.** The top ten water contaminants include lead, atrazine, PFAS, arsenic, GenX, hexavalent chromium, chloramines, fracking chemicals, TCE, and microplastics.[116] Learn more at EWG's tap water atlas or sign up for *The Brockovich Report*.
- **Skip bottled water and drink filtered tap instead.** A study found thirty-eight pollutants—including disinfection byproducts, radioactive chemicals, and industrial chemicals—in ten water bottle brands. The carbon footprint of bottled water is *3,500 times* worse than tap water.

 Check out *Consumer Reports* Guide to Water Filters to find what filter will work best for you.[117]

# Green School Lunch

The national school lunch program costs $15 billion annually and serves meals to nearly 30 million kids a day during the school year. If we green school lunch in the United States, we can reduce carbon emissions to the equivalent of taking 150,000 cars off the road each year. That's why greening the supply chain of school food can transport us to a healthier, more climate-friendly future.[118]

The vast majority of kids in the United States consume most of their calories at school.[119] The school lunch program is a critical lifeline for families and for children's health, but *what* food gets served has always been a hot-button issue. You may recall the 1980s controversy when efforts to cut billions from the school lunch program made it possible for ketchup to be considered a vegetable.[120] Because the cost of the lunch outweighs federal financial support, local school systems close the gap by outsourcing to private contractors.

Even after updating its nutrition guidelines in 2015 and offering more nutritious meals at school, the USDA gave in to industry pressure and allowed french fries and pizza. In 2018, the USDA once again conceded to the industry and watered down its nutrition proposals to allow more sodium, flavored milk, and refined grains. While nutrition standards have improved, there's still much work ahead.[121]

# How to Green School Lunch

How our food is grown, our personal health, and climate change are interconnected. Here are some great examples of ways schools are addressing these needs:

- **Greening school kitchens.** The nonprofit Eat REAL (think a green, sustainably grown certification for school meals) works with nutritionists to reduce sugar; increase whole-food options; create healthier, more plant-based and planet-friendly choices; and reduce food waste at scale. In 2020, Eat REAL helped the Mt. Diablo School District in California eliminate more than ten pounds of sugar per student and ensure that over one-third of produce served was sourced locally.[122]
- **Connecting farmers to schools.** The USDA Farm to School program provides grants to introduce students to local farmers and understand where their food comes from.[123] FoodCorps enlists young people in service to work for school food administrators. Soul Fire Farm trains activists in Afro-indigenous farming and food justice advocacy and has sponsored more than one hundred community gardens.
- **Scratch cooking** in school kitchens, rather than using prepackaged food, is becoming more popular. For example, the Chef Ann Foundation provides a grant program to train food service systems in scratch cooking and reaches 75,000 students through twenty-one school districts.[124]

- **School garden programs** like Edible Schoolyards and the Captain Planet Foundation's Project Learning Gardens report that participants have more interest in cooking at home and in environmental stewardship after gardening.

What was the cafeteria food like in your school? What worked? What didn't?

Support children's health through advocating for free school meals for all public school students by writing, calling, or emailing members of Congress. Call 202-224-3121 to reach the Congressional switchboard or look up your member of Congress here:

# DAY 47
# Check Your Household Goods

Study after study shows that toxic chemicals used in the manufacture of household goods like stain removers, upholstery fabric treatments, and other cleaners end up in our bodies.[125]

Here's an overview of some categories of household goods you should reexamine:

**Flame retardants.** Despite the phaseout of some classes of toxic flame retardants, they are still found everywhere. Organophosphates have largely replaced them but they share the same environmental concerns. Avoid polyurethane foam and consider using wool or mattresses with natural materials. Buy close-fitting children's PJs made from organic cotton.

**Nonstick chemicals.** These PFAS, more commonly known as "forever chemicals," are used in nonstick coatings, in food packaging, and to treat carpets and fabrics to make them stain resistant. This class of chemicals has been linked to obesity and heart disease. Avoid treated clothing, like rain jackets and all-weather wear. Consider checking your favorite outdoor brands on bluesign.com, which monitors companies' commitments to PFAS phaseouts.[126] Bluesign certifies that items carrying its official label have been manufactured to strict safety and environmental requirements. Cook in a cast-iron skillet to avoid unnecessary exposure to PFAS.

**Phthalates in clothing.** These chemicals, found in cosmetics, plastics, and food packaging, are also used in clothing processes. Exposures to phthalates have been linked to hormone disruption, early onset of puberty, attention deficit disorder, diabetes, and lower IQs. Phthalates in fashion can also cause skin irritation, and children are especially susceptible.[127] These chemicals are hard to avoid, but choose natural fabrics when you can and wash your kids' new clothes before they wear them.

To find non-toxic sustainable household goods, turn to the nonprofit certifier Made Safe.

To find PFA-free clothing, check out the Green Policy Institute's PFAS Central site.

# DAY 48
# Celebrate the National Parks

I never used to think about how my actions today might affect my kids and my grandkids, but I do all the time now," said Dan Wenk, former superintendent of Yellowstone National Park. Dan served in the National Park Service for forty-one years and says that for significant climate policy action to happen, "education is incredibly important because people have to accept there's a problem before they change their behavior."

Dan spent his youth exploring woods and streams around his house. "I wanted to go into the design field, and landscape architecture seemed like the right place to be. That interest led me to the National Park Service [NPS]." The built environment of the national parks, called *parkitecture*, still inspires him.

The mission of NPS is to preserve unimpaired natural and cultural resources and values of the National Park System for the enjoyment, education, and inspiration of this and future generations. The Park Service cooperates with partners to extend the benefits of natural and cultural resource conservation and outdoor recreation throughout this country and the world.

As a bureau within the Department of Interior, "NPS doesn't regulate as much as educate," said Dan. "One of the ways we could educate best was not by hitting people over the head with 'this is what you need to do,' but by showing them great examples of environmentally friendly architecture and landscapes. NPS has an audience of 300 million people a year. You can see how we do things and apply them to your daily life, at your home, or in your community."

When asked about the climate crisis, Dan remarked that "*depressing* is an apt word to describe it. You don't have to look very far to see the impacts of climate change. Yellowstone was established 150 years ago, so we have an incredible amount of data. I was in Yellowstone twice, from 1979 to 1984 and 2011 to 2018. By 2011 there were thirty less days of freezing temperatures a year. Snowpack was affected, which impacts river systems and water storage facilities. We've seen changes in wildlife migration, in fires and duration, and in habitat."

Dan remarked that the climate change trends are "undeniable," but that lawmakers tend to look at "episodes." "The problem is that we change our experiences when climate-induced extreme weather interrupts our life, when we need to change our actions."

To get educated on the different parks in the United States and learn how you can volunteer, check out:

Make plans to visit a park near you.

# DAY 49
# Be Wary of Forever Chemicals

In Parkersburg, West Virginia, attorney Rob Bilott fought a twenty-year legal battle against chemical manufacturer DuPont.

He uncovered corporate malfeasance and exposed a new class of dangerous, persistent chemicals, called "forever chemicals," that contaminate nearly every living thing on the planet.[128] (You can watch the exciting story in the movie *Dark Waters*, starring Mark Ruffalo.)

DuPont released toxic chemicals related to its nonstick line into the water, air, and nearby landfills. Company scientists monitored workers and found evidence of chemicals not breaking down and even passing through the placenta into the umbilical cord blood of newborns. DuPont researchers found correlations between exposure and the following health effects: cancer, thyroid dysfunction, pancreas and liver issues, colitis, and hormone disruption. DuPont quietly phased out the chemical but kept their essential health research secret.[129] In June 2023, Dupont entered into a $10.3 billion settlement with water utilities for polluting drinking water around the country with forever chemicals.[130]

The EPA uses the term "PFAS" (perfluorinated chemicals) to describe these forever chemicals, which include more than nine thousand substances. These chemicals are used in fire-fighting foam, pizza boxes, fast-food packaging, and nonstick cookware. These nonstick chemicals pollute the air, water, soil, human and animal fat, and human blood. Studies show that 99 percent of Americans have forever chemicals in their bodies, and they likely pollute the drinking water of the majority of Americans, more than two hundred million people.[131] At least twenty-four states have taken action on forever chemicals as of August 2023, and pressure continues to mount for strong federal regulation.[132]

In March 2023, for the first time in twenty-five years, the EPA made the historic decision to recommend that a group of chemical contaminants be regulated under the Safe Drinking Water Act. You guessed it, they finally took action and chose these forever chemicals. Note that EPA set a maximum contaminant level (the pollution limit) of 0.3 parts per trillion. To get a sense of how small that is, visualize it as the equivalent of finding a grain of sugar in an Olympic sized pool. The public health advisory EPA issued in 2023 set the standard at .004 parts per trillion—a quadrillionth—essentially declaring that there is no safe limit of forever chemical exposure in drinking water.[133] While EPA regulations move forward and cleanup continues, the best protection for consumers right now is to fight for reform and take steps to reduce your exposure.

Learn more about PFAS in drinking water here:

Learn more about how to try to avoid forever chemicals here:

## DAY 50
# Store Food the Right Way

- **Try to avoid plastic containers.** Use glass storage when
  you can. Plastic containers can leach toxic hormone-
  disrupting chemicals like BPA or BPA-related compounds
  and phthalates into the food. Even BPA-free containers may
  contain phthalates that can leach into food when used to
  reheat leftovers in the microwave.

- **Be mindful of expiration dates**. My kids love examining these
  dates and tossing anything that might be expired, but it's
  important to learn the difference between "use by" and
  "sell by" definitions. According to USDA, other than infant
  formula, dates don't indicate the safety of the food. For
  example, "best if used by" language indicates flavor, not the
  safety of the food. The term "sell by" is used for inventory
  management. "Freeze by" and "use by" are dates that refer
  to quality, not the safety of the food.

- **Think about food placement in your fridge.** The FDA
  encourages you to keep your fridge at 40 degrees or below
  to keep food safe. Place items you think might spoil soon at
  eye level, at the front, so you use them.

- **Use your freezer.** If you have leftovers, think about freezing
  them for the future. Be sure to include a sticker with the

date. This works great for meals like casseroles, lasagna, and soups.

- **Donate excess.** If it turns out you missed the mark with your menu plan and have leftover non-perishable goods, donate them to a local food bank. Hunger is on the rise in America, and food banks are always in need. Find your local food bank.

 Scan the QR code below to find your local food bank and learn how you can help.

# DAY 51
# Upgrade Your Travel

The transportation industry is a big carbon polluter. Consider taking a bike, walking, or carpooling when you can. Try out a bike rental or even an e-bike for short trips like groceries, etc. Be sure to talk to your tax advisor or research all the new tax incentives for hybrid and electric cars. Also consider buying offsets, but just like green cleaners, you want to research the companies to make sure they are for real.

Consumers may directly buy carbon offsets for trips or commutes, and it's pretty easy. At the end of each year, I even purchase offsets for my family—which on average produces forty-eight metric tons of carbon dioxide equivalent—for around $280. Carbon offset companies sell credits or investments in projects that create carbon sinks. Think reforestation or soil conservation projects. Given that the average cross-country flight is about 360 kilograms of carbon per person, or $7 to $10 per cross-country trip, being mindful about travel can reduce your footprint.[134] Many airlines offer ways for consumers to offset travel when purchasing tickets. Before you buy offsets, make sure the projects are verified by the EPA and beware of greenwashing. Ensure that the type of projects align with your values by researching the details.[135] Make sure the company you're supporting reports back on how it's using your investment.

 Think about your travel. Are there ways you can green your travel by walking, biking, carpooling, offsets, or going electric? Can you advocate for safe bike lanes and cleaner transportation in your community? Consider buying offsets for your next trip, but research them carefully.

# Connect Racial Justice & Sustainability

**M**ost of the toxic petrochemical plants in the United States are near or next to Black, Indigenous, Latinx, or communities of color. In addition, most of these communities do not have access to natural green spaces or to greener, safer products, which are sometimes not marketed to these consumers. As we embrace sustainability and a new green economy, equity and justice are paramount.[136]

In 2017 a study by the NAACP and the Clean Air Task Force showed that African American citizens are 75 percent more likely to live in a "fence-line" community that borders a toxic industrial facility. Big corporations and companies must understand the impact of their actions and minimize the harm they're creating.

The same goes with city planners. A 2021 study showed that extreme heat concentrates in predominantly BIPOC and poor areas due to less access to trees, more pavement, and more people. Even with the researchers blocked out data on income, counties with larger Black and Latino populations had significantly fewer areas with tree cover, parks, and natural cooling places than predominantly White neighborhoods.[137] How we design and locate green spaces affects communities too.

Andrea Ambriz knows this all too well. "I think people need to know that you can change systems and policies by showing up, doing excellent work, and getting involved in meaningful ways," explained Andrea. Named one of the "40 Under 40: Latinos in American Politics" by the *Huffington Post*, Andrea served as the head of External Affairs for the California Natural Resources Agency (CNRA).

Andrea approaches environmental issues through an intersectional lens of climate and economic justice. A former White House official, Andrea returned to California to serve the communities she first supported as a young staffer in the state legislature. "My initial work in community development brought me into environmental justice and conservation." She drafted an innovative grant program to invest in green spaces in underresourced communities in 2007. More than fifteen years later, Andrea's work still has ripple effects in pocket parks, playgrounds, and open green areas across the state.

To create a bold, sustainable future, she says "we must invest resources in communities hurt by environmental injustice to recover, repair, and protect them from climate change. Future generations also deserve that opportunity as a down payment toward building a new generation of community opportunity and wealth."

Talk to your loved ones about asking how products are made and where they're sourced. Talk about racial equity in sustainability and have the conversations that will create a perspective change. From planting trees & selling green brands in local stores to building energy efficient structures in BIPOC neighborhoods, we must be intentional to advance racial justice.

Read more about environmental justice and supporting fenceline communities:

## DAY 53
# Shop Smarter & Greener

You can make a difference. While some climate solutions require huge market and policy shifts, being mindful about the food you buy and use is one of the most powerful actions you can take to create a greener future. According to the nonprofit think tank Project Drawdown, individual and household actions could constitute 25 percent of the global carbon emission reductions we need to curb global warming. Learning how to reduce food waste is an essential strategy to becoming a better steward of the planet.

- **Embrace the ugly.** Fruit that is bruised, damaged, or just plain ugly can be used for soups, jams, smoothies, and grilling. It's estimated that one-third of all fruits and vegetables are deemed by farmers as too ugly to sell. So don't shy away from that ugly produce—it tastes the same, and you're doing your part to reduce food waste.
- **Create a shopping list.** This type of discipline keeps your eyes

from wandering at the cash register or splurging for the new, snazzy snack you see in your favorite aisle. Writing down a list and creating a plan helps save money too. If lists aren't your thing, you may want to take a "shelfie" of your pantry or fridge to remind you of what you already have at home.

- **Buy only what you need.** We all love big-box stores, but be conscious about buying in bulk. Make sure it's something you really need and not an 84-ounce bottle of a rare sauce you love but your family hates. Even if it's on sale, if it ends up in the landfill, it's not worth it.
- **Look for compostable packaging,** or at least skip the single-use plastic when you can—and especially avoid fruits and vegetables that are shrink-wrapped in plastic. Don't forget to bring your reusable shopping bags!
- **Look for low-mercury, sustainable fish.** Overfishing is a serious problem globally, and buying sustainably and using fish dishes efficiently is better for the environment as well. The Monterey Bay Aquarium publishes a helpful guide to sustainable fisheries.

To learn more about the role household actions have in solving climate change, scan the QR code to read an article from Project Drawdown.

To download the Monterey Bay Aquarium sustainable seafood guides, scan the QR code below.

## DAY 54
# Be a Booster for Environmental Education

Environmental education helps students achieve in the sciences and results in higher overall test scores and better critical thinking skills.[138] School systems across the country are investing in environmental and outdoor education, like in Oregon, Maryland, New Jersey, and California. Providing outdoor camp experiences, sponsoring hunting and fishing classes for kids, and bringing back recess and outdoor play as part of school are also key ways to connect children to nature.[139] Study after study shows that time outside makes students less stressed out and happier.

Take Christine Hill, known as "Chris," who never thought of herself as an outdoorsy person. Then one summer her parents enrolled her in an outdoor camp session they'd won in a school fundraiser. Instead of being miserable, Chris loved rock climbing, hiking, and camping. "I fell in love with nature, with rock climbing, with the people. And then I asked if I could go the full summer. I went back for six summers and ended up becoming an instructor."

Chris also learned about leadership and nature. "This outdoor camp was the most pivotal moment in my life." It helped her realize "what being in nature means to me, how it rejuvenates my soul, my mind, how it makes me feel good. If I want to spend time in nature and want my kids and grandkids to have this experience, what do I need to do to protect it?"

A star student, Chris graduated high school a year early and went to college at Appalachian State University. While studying abroad in Costa Rica, Chris worked with local lawyers and organizers to safeguard a small village from a planned gold mine. This sparked her interest in law and policy. She graduated from Vermont Law School.

Chris didn't want to be in a courtroom, but she wanted to change policy to center communities' needs. Now as Sierra Club's Chief Conservation Officer, Chris works to ensure equitable access to the outdoors for all people.

Despite being the subject of the award-winning Banff Film Festival documentary *Where I Belong*, which is about creating outdoor spaces for Black, Indigenous, and communities of color, Chris describes her leadership style as "behind the scenes." She aims to "lift up other people, push others forward, connect people, and especially lift up the next generation." What gives her hope is her belief in our shared humanity. "There are so many passionate people in this world, that may or may not have tapped into what it means to be stewards of the environment. But with just a little information they get it."

Have you had a meaningful environmental education experience either at a camp, on vacation, or in school? If so, where and how? If not, it's not too late to learn.

Find out more about environmental education in your community through the North American Association for Environmental Education.

## DAY 55
# Practice Self-Compassion & Resilience

Self-compassion serves to support overall well-being, address the eco-anxiety from the stressful events in our rapidly changing world, and prevent burnout. Psychologist Kristin Neff's research shows that self-compassion, which is basically treating yourself like you would a good friend, increases resilience. Neff's three elements of self-compassion are (1) self-kindness, (2) common humanity, and (3) mindfulness.[140]

Our inner critic doesn't result in motivation or efficiency. In fact, negative self-talk prevents us from fully appreciating success.

Neff urges patients to choose kindness instead of self-criticism, community connection over isolation, and to be present to observe what you're experiencing. Research on the health benefits of self-compassion reveals decreased levels of the stress hormone cortisol and increased heart rate variability, which means the variation in the time between heartbeats. (The higher the variation, the more relaxed you are. You can switch gears from relaxed to action faster.)[141] Self-reflection, meditation, and prayer can all increase feelings of happiness and lead to resilience. That feeling of resilience reduces burnout and energizes you.

At this point you might be wondering, *If this is a book about service to others and the planet, why are we supposed to think about ourselves?* First, because it's in society's best interests for you to be the best version of yourself. Second, people who are self-compassionate focus on self-improvement to do great work. Finally, Neff distinguishes *narcissism*, an unhealthy obsession and overestimation of one's importance, from *self-appreciation*, appreciating your strengths and recognizing that all people have goodness. Neff says that focusing on your strengths and talents is a way of "humbly honoring those who have helped us become the person we are today."[142] Taking care of ourselves and focusing on our unique talents both respects our ancestors and links us to future generations.

Reading the news about the climate crisis is anxiety-provoking, and the potential loss is mind-numbing. As your daily practice of sustainability evolves and you become more knowledgeable about the challenges we face, at times you'll need to rest and practice self-compassion.

Reflect on who has been compassionate to you in a meaningful way recently. How have you been self-compassionate this week?

## DAY 56
# See How Wildlife Is Linked to Human Health

Internationally, there is a substantial legal and illegal trade in wildlife. Millions of rare animals are smuggled across borders and sold. In the United States, which imports the most mammals and amphibians for pets, only a few animals crossing the borders are examined. As habitats are destroyed more animals crowd together in smaller spaces, which can result in "zoonosis" and make them more susceptible to disease.

Pandemic disease outbreaks, like coronavirus, and biodiversity are interlinked. According to the Centers for Disease Control, three out of four emerging viral diseases have spread from animals to humans. Called "zoonotic diseases," they transfer from animals to humans by direct contact or indirect contact like contaminated drinking water, food, air, or biting insects. The Brookings Institution concluded that we must "fundamentally revise our relationship with nature" to prevent more and more pandemics on a global scale. The wildlife trade—both legal and illegal—results in transmission of infectious disease. Creating more space for wildlife, curbing illegal wildlife trafficking, and preserving habitat are essential strategies to reduce zoonotic diseases.[143]

Preserving genetic diversity also translates into greater potential for medicinal plants for human health and nutrition. The National Wildlife Federation underscores that "fifty-six percent of the 150 most popular prescribed drugs are linked to discoveries of natural compounds, with an annual economic value of $80 billion." Think digitalis, which treats heart disease, the rosy periwinkle, which could help with cancer treatments, and the Pacific yew, whose compounds are already used in cancer treatments.[144]

Nature provides food for humans, and insects pollinate our plants. Nature supports life as well as our economy. There's a wonderful term for this obvious statement: "ecosystem services." This means that clean air, water, soil, and land are essential to our economies and how human society operates. Plants also provide building materials like wood and rubber. Framing nature as "services" is an effort to break through the old "jobs versus protecting the environment" and shift to a "protecting the environment equals jobs" paradigm. By embracing long-term thinking, we can inspire others to adopt a new worldview that without a healthy environment our economies will fail.

 Learn more about zoonotic diseases and how germs spread between animals and people here:

# Vote & Contact Your Elected Officials

**M**y middle school civics teacher used to say "Democracy is not a spectator sport." One of the most effective ways you can help create a greener, healthier, more just world is by voting and reaching out to your elected officials. Support candidates who reflect your values—those who not only believe climate change is real but will push back against the fossil fuel industry and its lobbyists. We need federal, state, and local officials to demand swift, comprehensive climate action and scale solutions. We also need a strong, independent judiciary to ensure that our rights to clean water, air, and energy are protected.

So make a plan to vote, volunteer or donate to a campaign, and show up at town halls and rallies. Join or sign up for the newsletter of a voting rights protection organization to ensure free and fair elections and full participation of citizens in our democracy.

Call your member of Congress and urge them to support strong climate action or a specific climate solution. The Congressional switchboard number is (202) 224-3121. Enter your zip code, and they'll connect you to your senator's or representative's office.

If you need help finding your representative's information, you can use the Find Your Representative feature on www.house.gov /representatives/find-your-representative. There is no central listing of member office public email addresses, but you can usually find a contact form (or email address) on the member's website.

If you don't speak up, your representatives aren't necessarily aware of the issues. Your elected officials can also provide you with valuable information on the legislation in process. Reminder: it's their job to talk to you.

Before reaching out, keep in mind key points:

- Make sure you know what you want to discuss. Consider writing your points down or typing them into your notes app for easy reference.
- Address the "5 Ws" (Who? What? Where? When? Why?).
- Mention and cite your resources to present facts and back up your opinions.
- Phone calls and emails work, but snail mail also has a big impact because it's becoming a rarer form of communication.

 For more guidance on contacting your member of Congress and talking points on range of green issues, check out NRDC's Action Center.

To learn more about how your member of Congress or state representative voted on environmental issues, check out the League of Conservation Voters Scorecard.

## DAY 58

# Check In with Yourself & Be Hopeful

---

"Some of my earliest memories relate to drought. We were not supposed to let the water run while we were brushing our teeth. We took short showers, not baths," recalled Ami Aronson, president of the Bernstein Family Foundation. "Protecting the environment was a way of life for us; I don't remember making a conscious decision about it." Ami's father was a college professor at University of California at San Francisco and taught environmental occupational health. "The connection between health and the environment was all around me, including in our dinner conversations," she continued.

Ami's work amplifies and invests in sustainability, the arts, and Jewish history and culture, and her leadership centers on empowering others. "Sure, I'm in" is a common refrain as she bolsters her personal mission and her influential and inspiring network of changemakers. Ami pointed out, "It's not just about investing in projects but showing up for your community."

Hope is part of Ami's confident, easygoing spirit. "I witness hope every day in the amazing people I meet. My nephew is headed to college in the fall and is working on technology to help monitor forest fires. My eighty-five-year-old friend recently moved to Israel to invest social capital in health technologies in underserved communities. A dear friend of mine was incarcerated for twenty-two years. As a spoken word and visual artist, his story and kind spirit inspire me. Hope is everywhere if you look for it," she remarked.

Ami understands that the climate crisis looms large. "You can be paralyzed by the enormity of the crisis, or you make a difference. If you are intentional about doing something each day, it'll be enough for your lifetime."

In climate action, Ami recommends that you start local with your impact. "I think so much of getting involved comes from proximity. Our country needs us. Start locally at your school. Attend a fundraiser. Read. Research. Find someone who inspires you."

Ami also insisted that "you don't have to reinvent the wheel" because so many organizations out there are doing amazing work. "Share your joy of action with someone else."

 Take a few minutes and write down or acknowledge aloud the hope that exists in your life.

Download the Joy Tracker to consider how you feel about your daily practice of sustainability and your progress.

## DAY 59
# Create Lasting Change

A multigenerational partnership is a major factor to create lasting change. Our experiences and anxiety about climate change may differ, but our need for each other and for action are constant. An Intergenerational Partnership sounds great, but how do we begin? In some cultures, intergenerational conversation and activities are the norm. Here I'm encouraging you to have an intentional discussion about *climate change* and the future.

This initial step requires you to spend time together either virtually or in person. This partnership doesn't have to be with your family; it can be with your community. What can you design or lay the groundwork for now that benefits the future? What behaviors can you change individually or together to support momentum for this vision?

As you collaborate, here are some Intergenerational Partnership principles to consider:

- **Ask.** The first step is to ask. Use open-ended questions about the issue—in this instance, climate change—and learn from those younger or older than you.

- **Listen.** This is a challenge for most of us. Consider setting a time limit and establishing respectful ground rules. Put your phone down. Don't interrupt or think about what you want to say next. For many young people, it's rare to be heard. Pause for two seconds before you respond. Be curious. Ask a follow-up question, and then ask another.

- **Share.** Talk about your own feelings, concerns, ideas. Don't assume the person you're talking with knows the historic event you're referring to or understands your perspective. Walk them through the experience you want to share.

- **Learn.** Commit to growth. This isn't easy, especially as we age. Hearing someone from a different generation explain their fears, dreams, and lessons learned can change your perspective.

- **Laugh.** This is easier said than done when you're talking about the climate crisis. If you're able to incorporate humor, even if it's laughing at yourself, you can open up space to be creative. Laughing is a way to break down intergenerational barriers and see the possibilities before you.

- **Brainstorm.** A partnership is not only listening and validating the other person's experience but also trying to form a positive vision of the future together. Ask "what if?" What needs to happen for your vision to become a reality? What can you do now to make an impact for the future?

- **Act.** After the previous six steps, come up with a plan. It doesn't have to be fancy, and you're not going to be graded on it. For example, you can choose one of the sixty greener initiatives in this book to do together. Promise to share articles on social media. Watch a documentary and then hop on Zoom to discuss it afterward.

# Discover Your Superpower

The most well-intentioned people often feel overwhelmed when they grasp the enormity of the climate crisis. But we can turn a sense of helplessness into a sense of accomplishment with small, consistent actions. Setting an intention each day to take a step to care for the planet can help ease our anxiety about the future, push the culture toward climate solutions, and create a sense of joy.

These routine habits can shift our collective consciousness to support comprehensive climate solutions. This realization motivated me to write this book, which is a call to action to identify your Service Superpower and create a daily sustainability practice to help protect the environment and our shared future. To inspire others to get involved, I also started a nonprofit organization, OneGreenThing, which tackles the mental health impacts of climate change.

The Service Superpower Assessment I developed after twenty years in the field of environmental advocacy identifies which of the seven essential service types best suits your personality. According to the Law of Identity, this identity match means you are more likely to maintain your new habits of sustainability. The assessment also provides a powerful way for you to discover how to contribute to the movement. The challenge for most of us is the basic but tough question: *How do I take the first step?*

Once you've identified your Service Superpower, you can continue a greener mindset and create a daily intention of action that best suits you.

Your superpowers, creativity, and unique talents are valued. Being part of this community can add more fun, compassion, and beauty to your daily life and create a brighter, more sustainable future. The compounding action of your sustainability habits and engaging your Service Superpower matters. Stay with it, because we're in an all-hands-on-deck moment for the planet. We still need to laugh. We still need joy. We still need hope. We can take the issue seriously without taking *ourselves* too seriously.

 Take the Service Superpower Assessment.

## VOTE FOR CLIMATE ACTION

Support local, state, and federal legislators that support strong clean energy and climate policies.

*Carbon emissions in the US saved by 2030:* **3 gigatons**

## REDUCE AIRLINE TRAVEL

Buy carbon offsets when you do.

*Carbon emissions saved:* **700 - 2800 kg/year**

## GREEN YOUR PORTFOLIO

Check your retirement fund or savings account. Consider investing in renewables and divest from fossil fuels.

*Carbon emissions saved:* **A LOT!**
*(AND $14 trillion us in the divest movement)*

## EAT MORE PLANTS, LESS MEAT

Reducing your meat intake by half can cut your carbon footprint by 40%.

*Carbon emissions saved:* **300 - 1600 kg/year**

## WALK, BIKE, RIDE

Greening your commute by taking public transit or buying a more efficient car can also significantly reduce your footprint.

*Carbon emissions saved living "car-free":* **1000 - 5300 kg/year**

*Carbon emissions saved with a more efficient car:* **1190 kg/year**

*Sources: David Suzuki Foundation, "Top 10 Things You Can Do About Climate Change (July 2021), Seth Wynes & Kimberly A. Nicholas, "The climate mitigation gap: education and government recommendations miss the most effective*

## REDUCE FOOD WASTE

Most families throw out 3 kg of otherwise edible food a week. Eliminating food waste or composting can make a big impact.

*Carbon emissions saved:* 530 kg/year

## SWITCH TO GREEN POWER

Call your energy provider to see if you can make the switch.

*Carbon emissions saved:* 1000 - 2500 kg/year

## FORGET FAST FASHION

The fashion industry contributes 10% of global carbon emissions. According to the ThredUp fashion calculator, the average person's fashion footprint is 734.8 kg/year.

*Carbon emissions saved:* 440 kg/year

## WASH IN COLD WATER

Washing laundry in cold water can result in significant energy savings.

*Carbon emissions saved:* 250 kg/year

## TALK ABOUT THE CLIMATE CRISIS

The biggest impact you can make is with your family, friends, and community. Talk abou the need for big climate solutions. Reach out with compassion. Talk about being a good ancestor. Create a greener, heathier, more equitable world for the next generation.

*Saved:* our shared future

*individual actions, 2017 Environmental Research Letters 12; Paul Hawken, Drawdown: The Most Comprehensive Plan Ever Proposed to Reverse Global Warming. New York, New York: Penguin Books, 2017. Thred Up Fashion Calculator.*

# Glossary

**Bioaccumulation** the build-up of toxic substances in a living organism. At each level of the food chain, the toxic chemicals remain, making the top predator most at risk.

**Blue carbon** ocean and coastal vegetation that stores carbon.

**Circular economy** is a term used to describe a model of production and consumption that involves sharing, leasing, repairing, refurbishing, and recycling existing materials as long as possible.

**Clean energy** is energy that does not emit carbon pollution. For example, wind and solar energy.

**Climate change** refers to changes in temperature and weather patterns over a long-term period, primarily caused by human activities burning fossil fuels. Often used as a synonym for "global warming."

**Climate justice** a concept that addresses the just division, fair sharing, and equitable distribution of the burdens of climate change and its mitigation and responsibilities to deal with climate change. Also called "environmental justice."

**Climate resilience** the concept of coping with, adapting to, and stopping further global warming. Climate resilient technologies include using nature as a climate solution, designing climate-resilient buildings and roads, and being prepared for natural disasters.

**Conventional agriculture** the prevailing way of farming that focuses on maximum productivity and efficiency of crop yields with large investments in equipment, single crops or "monoculture," and extensive use of synthetic fertilizers, toxic pesticides, and herbicides.

**Desertification** the process of an overall decline in land productivity. Lands in dry areas of the world are turning into deserts as trees and soil are stripped away by unsustainable farming like practices resulting in soil erosion, overgrazing, and faulty irrigation practices.

**Eco-anxiety** also known as climate anxiety. Intense emotional stress about climate change.

**Endocrine-disrupting chemicals (EDCs)** substances in the environment (air, soil, or water supply), food sources, personal care products, and manufactured products that act like hormones in your body and interfere with normal function your body's endocrine system.

**Forever chemicals or perfluorinated chemicals (PFAS)** this class of chemicals never break down within the natural environment. Used in products that are resistant to stains, heat, and water, such as fast-food packaging, fire-fighting foam, stain removers, nonstick cookware, and pizza boxes. Low-dose exposure linked to birth defects, cancer, weakened immune systems, and reproductive problems.

**Genetically modified organisms (GMOs)** containing DNA that has been altered using genetic engineering. Note a new federal food labeling law uses the term "bioengineered" instead.

**Global warming** the increase in temperature of the earth's atmosphere from the greenhouse effect, where the sun's rays continue to heat the earth as heavy greenhouse gases act as a blanket and trap other lighter gases from escaping. This effect is caused by human activity, primarily from burning fossil fuels. Now often a synonym for "climate change."

**Land conservation** is the long-term protection and management of unused or underused land resources.

**Microgrid** a small, independently controlled power system that can be operated with, or apart from, the local distribution and transmission systems. Microgrids can be powered by renewable energy and operate off the grid.

**Offset** a purchase of a carbon pollution reduction program to offset another polluting activity. For example, buying an investment in a new soil conservation project to offset emission from a typical construction project.

**Organic** means that the food is grown without hormones, toxic pesticides, synthetic fertilizers, radiation, or genetically modified or bioengineered organisms (GMOs).

**Paris Agreement** a global treaty rated by 197 countries, including the United States in 2021, that aims to reduce carbon pollution to net zero no later than 2050 by limiting temperature rise by 1.5 degrees Celsius.

**Persistent Organic Pollutants (POPs)** chemicals that never break down and persist in nature and our bodies. The dirty dozen POPs include X, Y, and Z.

**Phthalates** a hormone-disrupting chemical used to make plastic stronger and more durable and as a stabilizer in perfumes. Found in vinyl flooring, plastic packaging, vinyl toys, and personal care products like shampoos, nail polish, perfumes, and aftershave lotions. Linked to reproductive issues, allergies, and asthma.

**Regenerative agriculture** a holistic way of managing land that focuses on soil health (its nutrients and ability to retain water) and balance with nature. Practices include crop rotation, cover crops, no-till farming, composting, agroforestry, and reduced use of toxic chemicals.

**Solar garden** centrally located solar photovoltaic (PV) systems that provide electricity to participating subscribers.

**Sustainable fashion** clothing that is designed, manufactured, distributed, and used in ways that are environmentally friendly.

**Watershed** an area of land that channels rainfall and snowmelt to creeks, streams, and rivers, and eventually to outflow points such as reservoirs, bays, and the ocean.

**Zoonosis** another word for "zoonotic disease," where a disease transfers from animals to humans by direct contact or indirect contact like contaminated drinking water, food, air, or biting insects. Examples include rabies, malaria, and the coronavirus.

# Notes

1. Jonathan Watts, "We Have 12 Years to Limit Climate Change Catastrophe, Warns UN," *The Guardian*, October 8, 2018, https://www.theguardian.com /environment/2018/oct/08/global-warming-must-not-exceed-15c-warns -landmark-un-report.
2. Vicky McKeever, "Nearly Half of Young People Worldwide Say Climate Change Anxiety Is Affecting Their Daily Life," CNBC, September 14, 2021, https://www.cnbc.com/2021/09/14/young-people-say-climate-anxiety-is -affecting-their-daily-life.html.
3. Asha C. Gilbert, "Climate Change, Racism and Social Justice Concerns Affecting Gen Z's Physical and Mental Health," *USA Today*, April 21, 2021, https://www.usatoday.com/story/life/2021/04/21/climate-change-racism -and-social-justice-major-concerns-gen-z/7289512002/.
4. United Nations, "Nations Agree to End Plastic Pollution," accessed October 17, 2023, https://www.un.org/en/climatechange/nations-agree-end -plastic-pollution.
5. Anne Lamott, "12 Truths I Learned from Life and Writing," TED2017, TED, April 2017, https://www.ted.com/talks/anne_lamott_12_truths_i _learned_from_life_and_writing/up-next?referrer=playlist-incredibly _soothing_ted_talks.
6. Calum Neill, Janelle Gerard, and Katherine D. Arbuthnott, "Nature Contact and Mood Benefits: Contact Duration and Mood Type," *Journal of Positive Psychology* 14, no. 6 (December 2018): 756–67, https://www.tandfonline .com/doi/citedby/10.1080/17439760.2018.1557242?s; Florence Williams, *The Nature Fix: Why Nature Makes Us Happier, Healthier, and More Creative* (New York: W. W. Norton, 2017), 25–31; Magdalena M. H. E. van den Berg et al., "Autonomic Nervous System Responses to Viewing Green and Built Settings: Differentiating Between Sympathetic and Parasympathetic Activity," *International Journal of Environmental Research and Public Health* 12, no. 12 (December 2015): 15860–74, https://www.ncbi.nlm.nih.gov/pmc /articles/PMC4690962/; Gwen Dewar, "12 Benefits of Outdoor Play (and Tips for Helping Kids Reap These Benefits)," Parenting Science, 2019, https://parentingscience.com/benefits-of-outdoor-play/.
7. Hope Reese, "How a Bit of Awe Can Improve Your Health," *New York Times*, January 3, 2023, https://www.nytimes.com/2023/01/03/well/live/awe -wonder-dacher-keltner.html.

8. Susan Clayton and Christie Manning, "Climate Change's Toll on Mental Health," American Psychological Association, March 29, 2017, https://www.apa.org/news/press/releases/2017/03/climate-mental-health; Sarah Jaquette Ray, *A Field Guide to Climate Anxiety* (Oakland: University of California Press, 2020).

9. "Any Anxiety Disorder," National Institute of Mental Health, accessed October 12, 2021, https://www.nimh.nih.gov/health/statistics/any-anxiety -disorder.shtml.

10. U.S. Department of Health and Human Services, Surgeon General, "Youth Mental Health Advisory," June 2021, https://www.hhs.gov/surgeongeneral /priorities/youth-mental-health/index.html.

11. Cigna, "The Loneliness Epidemic Persists: A Post-Pandemic Look at the Statue of Loneliness among U.S. adults," March 2022, https://newsroom .thecignagroup.com/loneliness-epidemic-persists-post-pandemic-look.

12. Sarah Kaplan and Emily Guskin, "Most American Teens Are Frightened by Climate Change, Poll Finds, and About 1 in 4 Are Taking Action," *Washington Post*, September 16, 2019, https://www.washingtonpost.com/science/most -american-teens-are-frightened-by-climate-change-poll-finds-and-about-1-in -4-are-taking-action/2019/09/15/1936da1c-d639-11e9-9610-fb56c5522e1c _story.html; John Zogby Strategies, *A Survey of Gen Z & Millennials' Behavior & Values*, United States Conference of Mayors (Washington, DC: US Conference of Mayors, 2020), 5, https://www.usmayors.org/wp-content /uploads/2020/01/USCM_National-Youth-Poll-FINAL.pdf.

13. Kate Julian, "What Happened to American Childhood," *The Atlantic*, May 2020, https://www.theatlantic.com/magazine/archive/2020/05 /childhood-in-an-anxious-age/609079/.

14. Tess Riley, "Just 100 Companies Responsible for 71% of Global Emissions, Study Says," *The Guardian*, July 10, 2017, https://www.theguardian.com /sustainable-business/2017/jul/10/100-fossil-fuel-companies-investors -responsible-71-global-emissions-cdp-study-climate-change; "Our Planet Is Drowning in Plastic—It's Time for Change!," United Nations Environment Program, July 2020, https://www.unep.org/interactive/beat-plastic -pollution/.

15. James Clear, *Atomic Habits: An Easy & Proven Way to Build Good Habits & Break Bad Ones* (New York: Avery, 2018), 17.

16. Clear, *Atomic Habits*, vii–viii.

17. Charles Duhigg, *The Power of Habit: Why We Do What We Do in Life and Business* (New York: Random House, 2014), 19–25.

18. "The Greenhouse Effect," UCAR Center for Science Education, accessed October 11, 2021, https://scied.ucar.edu/learning-zone/how-climate-works /greenhouse-effect.

19. Holly Shaftel, ed., "Climate Change: How Do We Know?," NASA: Global Climate Change, updated September 28, 2021, https://climate.nasa.gov/evidence/.

20. Mark Lynas, Bejamin Z. Houlton, and Simon Perry, "Greater than 99% Consensus on Human Caused Climate Change in the Peer-Reviewed Scientific Literature," *Environmental Research Letters* 16, no. 11 (October 2021), https://doi.org/10.1088/1748-9326/ac2966.

21. "Headline Statements from the Summary for Policyholders," Intergovernmental Panel on Climate Change, August 9, 2021, PDF, 1, https://www.ipcc.ch/report/ar6/wg1/downloads/report/IPCC_AR6_WGI_Headline_Statements.pdf; Brad Plumer and Henry Fountain, "A Hotter Future Is Certain, Climate Panel Warns. But How Hot Is Up to Us," *New York Times*, August 9, 2021, https://www.nytimes.com/2021/08/09/climate/climate-change-report-ipcc-un.html.

22. Jennifer Marlon et al., "Yale Climate Opinion Maps 2021," Yale Program on Climate Change Communication, February 23, 2022, https://climatecommunication.yale.edu/visualizations-data/ycom-us/; see also Jonathan Watts, "Case Closed: 99.9% of Scientists Agree Climate Emergency Caused by Humans," *The Guardian*, October 19, 2021, https://www.theguardian.com/environment/2021/oct/19/case-closed-999-of-scientists-agree-climate-emergency-caused-by-humans.

23. John Zogby Strategies, *A Survey of Gen Z & Millennials' Behavior & Values.*

24. Gregg Sparkman et al, "Americans Experience A False Social Reality By Underestimating Popular Climate Policies by Nearly Half," *Nature Communications* 13, August 23, 2022, https://www.nature.com/articles/s41467-022-32412-y; Matthew Ballew et al, "Americans Underestimate How Many Others in the U.S. Think Global Warming is Happening," Yale Climate Change Communication, July 2, 2019, https://climatecommunication.yale.edu/publications/americans-underestimate-how-many-others-in-the-u-s-think-global-warming-is-happening/.

25. Rebeeca Solnit, "Big Oil Coined "Carbon Footprint" to Blame Us for Their Greed. Keep Them on the Hook," The Guardian, August 23, 2021, https://www.theguardian.com/commentisfree/2021/aug/23/big-oil-coined-carbon-footprints-to-blame-us-for-their-greed-keep-them-on-the-hook.

26. "6 Arguments to Refute Your Climate-Denying Relatives This Holiday," Earth Day, updated December 20, 2021, https://www.earthday.org/6-arguments-to-refute-your-climate-denying-relatives/.

27. Bruce Lieberman, "1.5 to 2 Degrees Celsius of Additional Global Warming: Does It Make a Difference?," Yale Climate Connections, August 4, 2021, https://yaleclimateconnections.org/2021/08/1-5-or-2-degrees-celsius-of-additional-global-warming-does-it-make-a-difference/; "How Do We Know

That Humans Are the Major Cause of Global Warming?," Union of Concerned Scientists, updated January 21, 2021, https://www.ucsusa.org/resources/are-humans-major-cause-global-warming.

28. Benjamin Franta, "Early Oil Industry Disinformation on Global Warming," *Environmental Politics* 30, no. 4 (2021): 663–68, https://www.tandfonline.com/doi/pdf/10.1080/09644016.2020.1863703.

29. Benjamin Franta, "Global Warming: From Scientific Warning to Corporate Casualty," TEDx Talks, July 20, 2021, YouTube video, 12:06, https://www.youtube.com/watch?v=Mp1JGqp7YMI.

30. Neela Banerjee, Lisa Song, and David Hasemyer, "Exxon: The Road Not Taken," Inside Climate News, September 16, 2015, https://insideclimatenews.org/project/exxon-the-road-not-taken/.

31. Daniel Coughlin, "Americans are Drowning in Stuff. Here's Why,' msn.com, December 7, 2022, https://www.msn.com/en-us/news/us/americans-are-drowning-in-stuff-here-s-why/ss-AA151r2k#image=18; Mary Macvean, "For Many People Gathering Possessions is Just the Stuff of Life," Los Angeles Times, March 14, 2014, https://www.latimes.com/health/la-xpm-2014-mar-21-la-he-keeping-stuff-20140322-story.html; OfferUp Press Release, "Americans Have Too Much Stuff and Not Enough Money, Study Finds," PR Newswire, August 29, 2016, https://www.prnewswire.com/news-releases/americans-have-too-many-things-and-not-enough-money-study-finds-300319019.html; The Story of Stuff Fact Sheet 2020, https://www.storyofstuff.org/wp-content/uploads/2020/01/StoryofStuff_FactSheet.pdf.

32. UN Climate Change, "UN Helps Fashion Industry Shift to Low Carbon," United Nations Framework Convention on Climate Change, September 6, 2018, https://unfccc.int/news/un-helps-fashion-industry-shift-to-low-carbon; Christine Ro, "Can Fashion Ever Be Sustainable?," BBC Future, March 10, 2020, https://www.bbc.com/future/article/20200310-sustainable-fashion-how-to-buy-clothes-good-for-the-climate.

33. "How Much Do Our Wardrobes Cost to the Environment?," World Bank, September 23, 2019, https://www.worldbank.org/en/news/feature/2019/09/23/costo-moda-medio-ambiente.

34. "How Much Do Our Wardrobes Cost to the Environment?"

35. "Circular Economy: Definition, Importance and Benefits," European Parliament, updated May 24, 2023, https://www.europarl.europa.eu/news/en/headlines/economy/20151201STO05603/circular-economy-definition-importance-and-benefits.

36. "The Climate Crisis—A Race We Can Win," United Nations, accessed October 12, 2021, https://www.un.org/en/un75/climate-crisis-race-we-can-win.

37. Nathaniel Bullard, "Clean Energy Investment Sets $1.1 Trillion Record, Matching Fossil Fuels for the First Time, Time, January 26, 2023, https://time.com/6250469/clean-energy-investment-sets-1-1-trillion-record-matching-fossil-fuels-for-the-first-time/.

38. David Roberts, "How to Drive Fossil Fuels Out of the US Economy, Quickly," Vox, August 6, 2020, https://www.vox.com/energy-and-environment/21349200/climate-change-fossil-fuels-rewiring-america-electrify.

39. Chelsea Eakin, "The US Can Reach 90 Percent Clean Electricity by 2035, Dependably and Without Increasing Consumer Bills," Berkeley Public Policy, June 9, 2020, https://gspp.berkeley.edu/faculty-and-impact/news/recent-news/the-us-can-reach-90-percent-clean-electricity-by-2035-dependably-and-without-increasing-consumer-bills.

40. Paul Hawken, *Drawdown: The Most Comprehensive Plan Ever Proposed to Reverse Global Warming* (New York: Penguin Books, 2017), 220–23.

41. Corinne Le Quéré et al., "Temporary Reduction in Daily Global CO2 Emissions During the COVID-19 Forced Confinement," *Nature Climate Change* 10 (2020): 647–53, https://www.nature.com/articles/s41558-020-0797-x; "After Steep Drop in 2020, Global Carbon Dioxide Emissions Have Rebounded Strongly," IEA, March 2, 2021, https://www.iea.org/news/after-steep-drop-in-early-2020-global-carbon-dioxide-emissions-have-rebounded-strongly.

42. Justin Badlam and Jared Cox, "The Inflation Reduction Act: Here's What's In it, " McKinsey & Company, October 24, 2022, https://www.mckinsey.com/industries/public-sector/our-insights/the-inflation-reduction-act-heres-whats-in-it.

43. "Plant-Rich Diets," Project Drawdown; Sarah Taber, "Farms Aren't Tossing Perfectly Good Produce. You Are," *Washington Post*, March 8, 2019, https://www.washingtonpost.com/news/posteverything/wp/2019/03/08/feature/farms-arent-tossing-perfectly-good-produce-you-are/; "Explore Solutions to Food Waste," ReFED, accessed October 18, 2021, https://insights-engine.refed.com/solution-database?dataView=total&indicator=us-dollars-profit.

44. Christopher Flavelle, "Climate Change Tied to Pregnancy Risks, Affecting Black Mothers Most," *New York Times*, June 18, 2020, https://www.nytimes.com/2020/06/18/climate/climate-change-pregnancy-study.html.

45. Sarah Kaplan, "Climate Change Is Also a Racial Justice Problem," *Washington Post*, June 29, 2020, https://www.washingtonpost.com/climate-solutions/2020/06/29/climate-change-racism/.

46. Renee Cho, "Why Climate Change Is an Environmental Justice Issue," *State of the Planet*, Columbia Climate School, September 22, 2020, https://news.climate.columbia.edu/2020/09/22/climate-change-environmental-justice/;

Nicola Jones, "How Native Tribes Are Taking the Lead on Planning for Climate Change," Yale Environment 360, February 11, 2020, https://e360.yale.edu /features/how-native-tribes-are-taking-the-lead-on-planning-for-climate-change; Daisy Simmons, "What Is 'Climate Justice'?," Yale Climate Connections, July 29, 2020, https://yaleclimateconnections.org/2020/07/ what-is-climate-justice/.

47. Clear, *Atomic Habits*, 39.

48. BOTWC Staff, "50 Years Ago, Shirley Chisholm Was Sworn In as the First Black Congresswoman," Because of Them We Can, January 4, 2019, https://www.becauseofthemwecan.com/blogs/botwc-firsts/50-years-ago -today-shirley-chisholm-was-sworn-in-as-the-first-african-american -congresswoman.

49. Nouran Salahieh, "More than a Third of the US Population, from the Midwest to the East Coast, Under Air Quality Alerts from Canadian Wildlife Smoke," CNN, June 28, 2023, https://www.msn.com/en-us/news /us/more-than-a-third-of-the-us-population-from-the-midwest-to-the -east-coast-under-air-quality-alerts-from-canadian-wildfire-smoke/ar -AA1d8uLt.

50. Wynne Armand, MD, "Air Pollution: How to Reduce Harm to Your Health, Harvard Medical School Harvard Health Publishing, August 13, 2021, https://www.health.harvard.edu/blog/air-pollution-how-to-reduce-harm -to-your-health-202108132567.

51. Edward O'Brien, "Fire Ecology Professor Says Ecosystem in 'Uncharted Territory,'" Montana Public Radio, June 23, 2021, https://www.mtpr.org /post/fire-ecology-professor-says-ecosystem-uncharted-territory; Bill Tripp, "Our Land Was Taken. But We Still Hold the Knowledge of How to Stop Mega-Fires," *The Guardian*, September 16, 2020, https://www.theguardian .com/commentisfree/2020/sep/16/california-wildfires-cultural-burns -indigenous-people.

52. "The Climate Denial Machine: How the Fossil Fuel Industry Blocks Climate Action," 'The Climate Reality Project, September 5, 2019, https://www .climaterealityproject.org/blog/climate-denial-machine-how-fossil-fuel -industry-blocks-climate-action.

53. Sarah Kaplan, "By 2050, There Will Be More Plastics than Fish in the World's Oceans, Study Says," *Washington Post*, January 20, 2016, https:// www.washingtonpost.com/news/morning-mix/wp/2016/01/20/by-2050 -there-will-be-more-plastic-than-fish-in-the-worlds-oceans-study-says/; Maanvi Singh, "It's Raining Plastic: Microscopic Fibers Fall from the Sky in Rocky Mountains," *The Guardian*, August 13, 2019, https://www .theguardian.com/us-news/2019/aug/12/raining-plastic-colorado-usgs -microplastics; Matthew Green, "'Punch in the Gut' as Scientists Find Micro

Plastic in Arctic Ice," Reuters, August 14, 2019, https://www.reuters.com /article/us-environment-arctic-plastic/punch-in-the-gut-as-scientists-find -micro-plastic-in-arctic-ice-idUSKCN1V41V2; Melissa Locker, "It's Snowing Plastic in the Arctic Now," *Fast Company*, August 15, 2019, https://www .fastcompany.com/90390654/its-snowing-plastic-in-the-arctic-now.

54. OECD, Plastic Pollution Is Growing Relentlessly as Waste Management and Recycling Fall Short, Says OECD, February 22, 2022, https://www .oecd.org/newsroom/plastic-pollution-is-growing-relentlessly-as-waste -management-and-recycling-fall-short.htm.

55. Doyle Rice, "Oh, Yuck! You're Eating About a Credit Card's Worth of Plastic Every Week," *USA Today*, June 13, 2019, https://www.usatoday.com /story/news/nation/2019/06/12/plastic-youre-eating-credit-cards-worth -plastic-each-week/1437150001/.

56. Congressional Research Service, "Federal Land Ownership: Overview and Data," February 21, 2020, https://sgp.fas.org/crs/misc/R42346.pdf.

57. David Treuer, "Return the National Parks to the Tribes," The Atlantic, May 2021, https://www.theatlantic.com/magazine/archive/2021/05 /return-the-national-parks-to-the-tribes/618395/.

58. Marguerite Holloway, "Your Children's Yellowstone Will Be Radically Different," *New York Times*, November 15, 2018, https://www.nytimes.com /interactive/2018/11/15/climate/yellowstone-global-warming.html.

59. "Feeling Powerless? Switch to Green Power," Earth Day Network, April 11, 2020, https://www.earthday.org/feeling-powerless-switch-to-green-power/.

60. "Find Green-E Certified," Green-E Certified, accessed October 12, 2021, https://www.green-e.org/certified-resources.

61. Brady Seals, "Reality Check: Gas Stoves are a Health and Climate Problem," Rocky Mountain Institute Blog, February 15, 2023, https://rmi.org /gas-stoves-health-climate-asthma-risk/.

62. "Genetically Modified Organisms," *National Geographic*, accessed October 12, 2021, https://www.nationalgeographic.org/encyclopedia /genetically-modified-organisms/.

63. "Genetically Modified Organisms," *National Geographic*.

64. Stacy Malkan, "Glyphosate: Cancer and Other Health Concerns," US Right to Know, October 13, 2023, https://usrtk.org/pesticides/glyphosate-health -concerns/.

65. Malkan, "Glyphosate Fact Sheet."

66. Miles McEvoy, "Organic 101: What the USDA Organic Label Means," US Department of Agriculture, March 22, 2012, https://www.usda.gov/ media/blog/2012/03/22/organic-101-what-usda-organic-label-means; Kendra Klein and Anna Lappé, "You Have Pesticides in Your Body. But an Organic Diet Can Reduce Them by 70%," *The Guardian*, August 11, 2020,

https://www.theguardian.com/environment/commentisfree/2020/aug/11/pesticide-danger-organic-food-roundup-study.

67. M. Shahbandeh, "Total Area of Land in United States Farms from 2000 to 2022 (in 1,000 Acres)," Statista, accessed October 18, 2023, https://www.statista.com/statistics/196104/total-area-of-land-in-farms-in-the-us-since-2000/.

68. "Children Interrupt BBC News Interview—BBC News," BBC News, March 10, 2017, YouTube video, 00:43, https://www.youtube.com/watch?v=Mh4f9AYRCZY&t=3s.

69. Allison Aubrey, "Happiness: It Really Is Contagious," National Public Radio, December 5, 2008, https://www.npr.org/templates/story/story.php?storyId=97831171.

70. Jessica Cerretani, "The Contagion of Happiness," *Harvard Medicine*, Summer 2011, https://hms.harvard.edu/magazine/science-emotion/contagion-happiness.

71. "Endangered Species Act by the Numbers," 3.

72. "What Is the Relationship Between Deforestation and Climate Change?," Rainforest Alliance, updated August 12, 2018, https://www.rainforest-alliance.org/insights/what-is-the-relationship-between-deforestation-and-climate-change/.

73. Xiya Liang et al., "Research Progress of Desertification and Its Prevention in Mongolia," *Sustainability* 13, no. 12 (June 17, 2021): 6861, https://www.mdpi.com/2071–1050/13/12/6861.

74. Judith Schwartz, "Soil as Carbon Storehouse: New Weapon in the Carbon Fight?," Yale Environment 360, March 4, 2014, https://e360.yale.edu/features/soil_as_carbon_storehouse_new_weapon_in_climate_fight.

75. NOAA, "What Is Blue Carbon?," National Oceanic and Atmospheric Administration, updated November 24, 2021, https://oceanservice.noaa.gov/facts/bluecarbon.html.

76. Capi Lynn, "A Desperate Rescue: A Father's Heartbreaking Attempt to Save His Family from a Raging Fire," *Salem Statesman Journal*, September 11, 2020, https://www.statesmanjournal.com/in-depth/news/2020/09/10/oregon-wildfires-santiam-fire-evacuations-leave-family-members-dead/5759101002; Alejandra Borunda, "The Science Connecting Wildfires to Climate Change," *National Geographic*, September 17, 2020, https://www.nationalgeographic.com/science/article/climate-change-increases-risk-fires-western-us.

77. "Son, Grandmother Die in Fire, Mother in Burn Center," GoFundMe.com, September 30, 2020, https://ie.gofundme.com/f/help-chris-and-angie-after-devastating-loss?utm_campaign=p_cp_url&utm_medium=os&utm_source=customer.

78. Karissa Kovner, "Persistent Organic Pollutants: A Global Issue, a Global Response," United States Environmental Protection Agency, accessed October 21, 2021, https://www.epa.gov/international-cooperation /persistent-organic-pollutants-global-issue-global-response.

79. Robert O. Wright and Rosalind J. Wright, "The Institute for Exposomic Research," Icahn School of Medicine at Mount Sinai, accessed October 19, 2021, https://icahn.mssm.edu/research/exposomic.

80. Theo Colborn, Dianne Dumanoski, and John Peterson Myers, *Our Stolen Future: Are We Threatening Our Fertility, Intelligence, and Survival? A Scientific Detective Story* (New York: Plume, 1997).

81. Linda Birnbaum, "State of the Science of Endocrine Disruptors," *Environmental Health Perspectives* 121, no. 4 (April 2013): a107, https:// www.ncbi.nlm.nih.gov/pmc/articles/PMC3620755/.

82. "EWG's Guide to Endocrine Disruptors: 8 Hormone-Altering Chemicals and How to Avoid Them," Environmental Working Group, October 28, 2013, http://www.ewg.org/research/dirty-dozen-list-endocrine-disruptors.

83. Manoj Kumar et al., "Environmental Endocrine-Disrupting Chemical Exposure: Role in Non-Communicable Diseases," *Frontiers in Public Health* 8 (September 24, 2020), https://www.frontiersin.org/articles/10.3389/fpubh .2020.553850/full; Pamela D. Noyes et al., "The Toxicology of Climate Change: Environmental Contaminants in a Warming World," *Environment International* 35, no. 6 (August 2009): 971–86, https://pubmed.ncbi.nlm .nih.gov/19375165/; Allison J. Crimmins et al., *The Impacts of Climate Change on Human Health in the United States: A Scientific Assessment* (Washington, DC: U.S. Global Change Research Program, 2016), 22, https://health2016 .globalchange.gov/low/ClimateHealth2016_FullReport_small.pdf.

84. "Household Chemical Products and Their Health Risk," Cleveland Clinic, May 24, 2018, https://my.clevelandclinic.org/health/articles/11397 -household-chemical-products-and-their-health-risk.

85. Cliff Weathers, "5 Toxic Household Products You Probably Use Every Day," Salon, June 22, 2015, https://www.salon.com/2015/06/22/5_toxic _household_products_you_probably_use_every_day/.

86. Annie Lowrey, "All That Performative Environmentalism Adds Up," *The Atlantic*, August 31, 2020, https://www.theatlantic.com/ideas/archive /2020/08/your-tote-bag-can-make-difference/615817/.

87. Roman Krznaric, *The Good Ancestor: A Radical Prescription for Long-Term Thinking* (New York: The Experiment, 2020).

88. Lisa Zaval, Ezra M. Markowitz, and Elke U. Weber, "How Will I Be Remembered? Conserving the Environment for the Sake of One's Legacy," *Psychological Science* 26, no. 2 (February 2015): 231–36, https://journals .sagepub.com/doi/10.1177/0956797614561266.

89. Michael Sanders and Sarah Smith, "Can Simple Prompts Increase Bequest Giving? Field Evidence from a Legal Call Centre," *Journal of Economic Behavior and Organization* 125 (May 2016): 179–91, https://www .sciencedirect.com/science/article/abs/pii/S0167268116000044.

90. Water Science School, "The Water in You: Water and the Human Body," May 22, 2019, https://www.usgs.gov/special-topic/water-science-school /science/water-you-water-and-human-body; NOAA, "Rivers and Streams," *National Geographic*, accessed October 18, 2021, https://www .nationalgeographic.org/topics/resource-library-rivers-and-streams/; "What Is a Watershed?," National Ocean Service, updated February 26, 2021, https://oceanservice.noaa.gov/facts/watershed.html.

91. Perry Beeman, "Des Moines River 'Essentially Unusable' for Drinking Water Due to Algae Toxins," *Iowa Capital Dispatch*, August 26, 2020, https:// iowacapitaldispatch.com/2020/08/26/des-moines-river-essentially -unusable-for-drinking-water-due-to-algae-toxins/.

92. "Canned Foods," Center for Environmental Health, December 31, 2021, https://ceh.org/products/canned-foods/; "Endocrine Disruptors," National Institute of Environmental Health Sciences, July 12, 2021, http://www. niehs.nih.gov/health/topics/agents/endocrine/.

93. George Citroner, "Food Industry's Switch to Non-BPA Linings Still Poses Health Risks," Health Line, July 25, 2019, https://www.healthline .com/health-news/common-chemicals-in-plastics-linked-to-childhood -obesity#Bisphenol-disrupts-the-bodys-metabolism.

94. Michael Hawthorne, "Chemical Companies, Big Tobacco, and the Toxic Products in Your Home," Tribune Watchdog: Playing with Fire: *Chicago Tribune*, May 6, 2012, http://media.apps.chicagotribune.com /flames/index.html; Liza Gross, "Flame Retardants in Consumer Products Linked to Health and Cognitive Problems," *Washington Post*, April 15, 2013, https://www.washingtonpost.com/national/health -science/flame-retardants-in-consumer-products-are-linked-to-health -and-cognitive-problems/2013/04/15/f5c7b2aa-8b34-11e2-9838 -d62f083ba93f_story.html.

95. Valerie J. Brown, "Metals in Lip Products—A Cause for Concern?," *Environmental Health Perspectives* 121, no. 6 (June 1, 2013): A196, http://dx.doi.org/10.1289/ehp.121-a196.

96. Sarah Yang, "Teen Girls See Big Drop in Chemical Exposure with Switch in Cosmetics," Berkeley News, March 7, 2016, http://news.berkeley.edu /2016/03/07/cosmetics-chemicals/.

97. Julianna Deardorff and Louise Greenspan, *The New Puberty: How to Navigate Early Development in Today's Girls* (New York: Rodale, 2015); Heather White, "Are You There, God? It's Me, Rebecca: Early Puberty a New Normal,"

*Ms. Magazine*, February 4, 2015, http://msmagazine.com/blog/2015/02/04
/are-you-there-god-its-me-rebecca-early-puberty-a-new-normal/.

98. See "Red List," Campaign for Safe Cosmetics, accessed October 20, 2021,
    https://www.safecosmetics.org/get-the-facts/chemicals-of-concern/red-list/;
    Sally Wadyka, "What You Need to Know About Sunscreen Ingredients,"
    Consumer Reports, updated May 22, 2019, https://www.consumerreports
    .org/sunscreens/what-you-need-to-know-about-sunscreen-ingredients/.

99. Katherine Derla, "Sunscreen Ingredient Threatens Marine Life: Here's
    How Oxybenzone Kills Coral Reefs," *Tech Times*, October 22, 2015,
    http://www.techtimes.com/articles/98181/20151022/sunscreen-ingredient
    -threatens-marine-life-heres-how-oxybenzone-kills-coral-reefs.htm; Caroline
    Picard, "What the Proposed New FDA Sunscreen Rules Could Mean for
    You," *Good Housekeeping*, June 10, 2020, https://www.goodhousekeeping
    .com/health/a26470685/fda-sunscreen-regulations/.

100. "History of Bison Management in Yellowstone," National Park Service,
     February 12, 2021, https://www.nps.gov/articles/bison-history-yellowstone.htm.

101. Wildlife Species Information, "American Buffalo (*Bison bison*)," US Fish and
     Wildlife Service, accessed October 20, 2021, https://www.fws.gov/species
     /species_accounts/bio_buff.html.

102. Wildlife Species Information, "American Buffalo (*Bison bison*)."

103. Deborah Blum, *The Poison Squad: One Chemist's Single-Minded Crusade
     for Food Safety at the Turn of the Twentieth Century* (New York: Random
     House, 2019); Richard Fisher, "How to Decode a Food Label," BBC Future,
     June 23, 2021, https://www.bbc.com/future/article/20210623-how-to
     -decode-a-food-label.

104. Jensen Jose, "Cutting the GRAS," Center for Science in the Public Interest,
     July 28, 2021, https://www.cspinet.org/news/blog/cutting-gras.

105. Jose, "Cutting the GRAS."

106. Robb Q. Telfer, "De-Extinction Counter," Habitat 2030, September 9, 2016,
     https://habitat2030.org/blog/extinction/.

107. Bill Chappell and Nathan Rott, "1 Million Animals and Plant Species
     Are at Risk of Extinction, U.N. Report Says," May 6, 2019, on *All Things
     Considered*, NPR, podcast, 3:40, https://www.npr.org/2019/05/06
     /720654249/1-million-animal-and-plant-species-face-extinction-risk-u-n
     -report-says.

108. John H. Cushman Jr., "After 'Silent Spring,' Industry Put Spin on All It
     Brewed," *New York Times*, March 26, 2001, https://www.nytimes.com
     /2001/03/26/us/after-silent-spring-industry-put-spin-on-all-it-brewed.html.

109. Corry Westbrook, "Endangered Species Act by the Numbers," National
     Wildlife Federation, PDF, 1, https://www.nwf.org/~/media/pdfs/wildlife
     /esabythenumbers.ashx; "Endangered Species," National Wildlife Foundation,

accessed October 18, 2021, https://www.nwf.org/Educational-Resources
/Wildlife-Guide/Understanding-Conservation/Endangered-Species.

110. "Plant-Rich Diets," Project Drawdown, accessed October 12, 2021,
https://drawdown.org/solutions/plant-rich-diets.

111. "Plant-Rich Diets," Project Drawdown.

112. "10 Common Climate-Damaging Foods," NRDC, infographic, accessed
October 12, 2021, https://www.nrdc.org/sites/default/files/10-common
-climate-damaging-foods-infographic.pdf.

113. Charles Duhigg, "That Tap Water Is Legal but May Be Unhealthy," *New York
Times*, December 16, 2009, https://www.nytimes.com/2009/12/17/us
/17water.html.

114. "Drinking Water Regulations," United States Environmental Protection
Agency, accessed October 19, 2021, https://www.epa.gov/dwreginfo/drinking
-water-regulations; Mae Wu, "The Safe Drinking Water Act Must Be Updated
Now," Natural Resources Defense Council, July 28, 2020, https://www
.nrdc.org/experts/mae-wu/safe-drinking-water-act-must-be-updated-now.

115. Wu, "Safe Drinking Water Act."

116. Erin Brockovich, *Superman's Not Coming: Our National Water Crisis and
What We the People Can Do About It* (New York: Pantheon, 2020), 61–98.

117. "Guide to Safe Tap Water and Water Filters," Food & Water Watch,
February 16, 2016, https://www.foodandwaterwatch.org/2016/02/16
/guide-to-safe-tap-water-and-water-filters/; "Five Reasons to Skip Bottled
Water," Environmental Working Group, September 22, 2013, https://
www.ewg.org/consumer-guides/five-reasons-skip-bottled-water; Joey
Grostern, "Environmental Impact of Bottled Water 'Up to 3,500 Times
Greater than Tap Water," *The Guardian*, August 5, 2021, https://www
.theguardian.com/environment/2021/aug/05/environmental-impact-of
-bottled-water-up-to-3500-times-greater-than-tap-water; "Water Filter
Ratings," Consumer Reports, accessed October 19, 2021, https://www
.consumerreports.org/cro/water-filters.htm.

118. Saied Toossi, "National School Lunch Program," US Department of
Agriculture, Economic Research Service, updated September 27, 2023,
https://www.ers.usda.gov/topics/food-nutrition-assistance/child-nutrition
-programs/national-school-lunch-program/; Kari Hamerschlag and
Christopher D. Cook, "How Greener School Lunches Can Help Fight Climate
Change," Green Schools National Network, April 6, 2017, https://
greenschoolsnationalnetwork.org/greener-school-lunches-can-help-fight
-climate-change/.

119. Centers for Disease Control and Prevention, "School Nutrition,"
February 15, 2021, https://www.cdc.gov/healthyschools/nutrition
/schoolnutrition.htm.

120. Amy Bentley, "Ketchup as a Vegetable: Condiments and the Politics of School Lunch in Reagan's America," *Gastronomica* 21, no. 1 (2021): 17–26, https://online.ucpress.edu/gastronomica/article/21/1/17/116213/Ketchup-as-a-VegetableCondiments-and-the-Politics.

121. Gaddis, "Big Business."

122. Jordan Shlain and Nora LaTorre, *Impact Report 2020* (Richmond, CA: Eat REAL, December 2020), https://hnanp3kj45p2kxeqv1pOsris-wpengine.netdna-ssl.com/wp-content/uploads/2020/12/Eat-REAL-Impact-Report-2020.pdf.

123. Gaddis, "Big Business."

124. "Our Impact," Chef Ann Foundation, accessed October 12, 2021, https://www.chefannfoundation.org/what-weve-done/our-impact.

125. *Fourth National Report on Human Exposure to Environmental Chemicals: Executive Summary* (Washington, DC: Centers for Disease Control and Prevention, 2009), PDF, https://www.cdc.gov/exposurereport/pdf/FourthReport_ExecutiveSummary.pdf; "Welcome to the Human Toxome Project," Environmental Working Group, accessed June 15, 2020, https://www.ewg.org/sites/humantoxome/.

126. Daniel Penny, "Is Your Beloved Outdoors Gear Bad for the Planet?," *GQ*, January 22, 2021, https://www.gq.com/story/outdoor-gear-pfas-study.

127. National Biomonitoring Program, "Phthalates Factsheet," Centers for Disease Control and Prevention, April 5, 2021, https://www.cdc.gov/biomonitoring/Phthalates_FactSheet.html; "Get the Facts: Phthalates," Safer Chemicals, Healthy Families, accessed December 31, 2021, https://saferchemicals.org/get-the-facts/toxic-chemicals/phthalates/; Joseph M. Braun, Sheela Sathyanarayana, and Russ Hauser, "Phthalate Exposure and Children's Health," *Current Opinions in Pediatrics* 25, no. 2 (April 2013): 247–54, https://www.ncbi.nlm.nih.gov/pmc/articles/PMC3747651/.

128. Nathaniel Rich, "The Lawyer Who Became DuPont's Worst Nightmare," *New York Times Magazine*, January 6, 2016, https://www.nytimes.com/2016/01/10/magazine/the-lawyer-who-became-duponts-worst-nightmare.html.

129. Nathaniel Rich, "The Lawyer Who Became DuPont's Worst Nightmare"; Cheryl Hogue, "DuPont, EPA Settle," *Chemical and Engineering News* 83, no. 51 (December 19, 2005), https://cen.acs.org/articles/83/i51/DuPont-EPA-Settle.html.

130. John Flesher, "3M reaches $10.3 Billion Settlement Over Contamination of Water Systems with Forever Chemicals, Associated Press, June 22, 2023, https://apnews.com/article/pfas-forever-chemicals-3m-drinking-water-81775af23d6aeae63533796b1a1d2cdb.

131. "PFAS Explained," US Environmental Protection Agency, accessed December 17, 2021, https://www.epa.gov/pfas/pfas-explained;

Christophe Haubursin and Mac Schneider, "How 'Forever Chemicals' Polluted America's Water," Vox, August 4, 2020, https://www.vox.com /videos/2020/8/4/21354034/pfas-forever-chemicals-water-north -carolina; Annie Sneed, "Forever Chemicals Are Widespread in U.S. Drinking Water," *Scientific American*, January 22, 2021, https://www .scientificamerican.com/article/forever-chemicals-are-widespread-in-u-s -drinking-water/.

132. Kimberly Kindy, "States Take Matters into Their Own Hands to Ban 'Forever Chemicals,'" Washington Post, June 5, 2023, https://www.washingtonpost .com/politics/2023/06/05/forever-chemicals-state-bans-pfas/.

133. "PFAS in Drinking Water: EPA Proposes Historic New Regulation," National Law Review (March 17, 2023), https://www.natlawreview.com/article /pfas-drinking-water-epa-proposes-historic-new-regulation

134. "Aviation," Carbon Independent, updated September 13, 2021, https:// www.carbonindependent.org/22.html.

135. Kiah Treece, "The 6 Best Carbon Offset Programs of 2022," Treehugger, updated March 19, 2021, https://www.treehugger.com/best-carbon -offset-programs-5076458.

136. Victoria Gilchrist and Heather White, "How to Talk About Racial Justice in Sustainability," GreenBiz, December 16, 2020, https://www.greenbiz.com /article/how-talk-about-racial-justice-sustainability.

137. Susanne Benz & Jennifer Burney, "Press Release: US-Wide, Non-White Neighborhoods are Hotter than White Ones," AGU Advancing Earth & Space Science, July 13, 2021, https://news.agu.org/press-release/us-wide-non -white-neighborhoods-are-hotter-than-white-ones/

138. ee Works, "The Benefits of Environmental Education for K-12 Students," North American Association for Environmental Education, accessed October 18, 2021, https://naaee.org/eepro/research/eeworks/student -outcomes.

139. Heather White, *Connecting Today's Kids with Nature: A Policy Action Plan* (Reston, VA: National Wildlife Federation, 2008), https://www.nwf.org /~/media/PDFs/Campus-Ecology/Reports/CKN_full_optimized.ashx.

140. Kristen Neff, "What Is Self-Compassion?," Self-Compassion, accessed July 1, 2021, https://self-compassion.org/the-three-elements-of-self -compassion-2/#3elements.

141. Kristen Neff, "The Physiology of Self-Compassion," Self-Compassion, accessed July 1, 2021, https://self-compassion.org/the-physiology-of-self -compassion/.

142. Kristen Neff, "Self-Appreciation: The Flipside of Self-Compassion," Self-Compassion, accessed July 1, 2012, https://self-compassion.org/self -appreciation-the-flip-side-of-self-compassion/.

143. One Health, "Zoonotic Diseases," Centers for Disease Control and Prevention, updated July 1, 2021, https://www.cdc.gov/onehealth/basics/zoonotic -diseases.html; Vanda Felbab-Brown, "Preventing Pandemics Through Biodiversity Conservation and Smart Wildlife Trade Regulation," Brookings Institution, January 25, 2021, https://www.brookings.edu/research /preventing-pandemics-through-biodiversity-conservation-and-smart-wildlife -trade-regulation/; Julie Shaw, "Why Is Biodiversity Important?,"Conservation International, updated May 17, 2021, https://www.conservation.org/blog /why-is-biodiversity-important.

144. Corry Westbrook, "Endangered Species Act by the Numbers," 3.